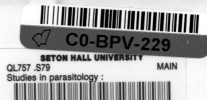

US-ISSN-0035-4996

RICE UNIVERSITY STUDIES

STUDIES IN PARASITOLOGY

In memory of
CLARK P. READ

J. E. BYRAM and GEORGE L. STEWART, Editors

C. ARME
HAROLD L. ASCH
J. E. BYRAM
KAY W. BYRAM
K. E. DIXON
SUE CARLISLE ERNST
JEANETTE C. ERTEL
FRANK M. FISHER, JR.
D. J. GRAFF
FRANCIS M. GRESS
C. A. HOPKINS
HADAR ISSEROFF

ARAXIE KILEJIAN
D. L. LEE
MICHAEL G. LEVY
RICHARD D. LUMSDEN
AUSTIN J. MacINNIS
JAMES S. McDANIEL
CLARK P. READ
FRANKLIN SOGANDARES-BERNAL
HELEN E. STALLARD
GEORGE L. STEWART
L. T. THREADGOLD
GARY L. UGLEM

Vol. 62, No. 4   Fall 1976

RICE UNIVERSITY STUDIES

published by

WILLIAM MARSH RICE UNIVERSITY

RICE UNIVERSITY STUDIES, successor to the RICE INSTITUTE PAMPHLET, is issued quarterly and contains writings in all scholarly disciplines by staff members and other persons associated with Rice University.

EDITOR: Katherine Fischer Drew

ASSISTANT EDITOR: Kathleen Much Murfin

RICE UNIVERSITY STUDIES REVIEW BOARD

K. F. Drew, Chairman, R. V. Butler, Margaret E. Eifler, Werner H. Kelber, and John E. Parish

INFORMATION FOR AUTHORS

Manuscripts submitted for publication should be addressed to the Editor, RICE UNIVERSITY STUDIES, Houston, Texas. If accepted, manuscripts will be edited to conform to the style of the STUDIES. Notes and references will appear at the end of the manuscript.

Individual numbers of RICE UNIVERSITY STUDIES may be purchased from the Rice Campus Store, P.O. Box 1892, Houston, Texas 77001.

Second class postage paid at Houston, Texas.

CLARK PHARES READ

RICE UNIVERSITY STUDIES

STUDIES IN PARASITOLOGY

In memory of
CLARK P. READ

J. E. BYRAM and GEORGE L. STEWART, Editors

C. ARME
HAROLD L. ASCH
J. E. BYRAM
KAY W. BYRAM
K. E. DIXON
SUE CARLISLE ERNST
JEANETTE C. ERTEL
FRANK M. FISHER, JR.
D. J. GRAFF
FRANCIS M. GRESS
C. A. HOPKINS
HADAR ISSEROFF

ARAXIE KILEJIAN
D. L. LEE
MICHAEL G. LEVY
RICHARD D. LUMSDEN
AUSTIN J. MacINNIS
JAMES S. McDANIEL
CLARK P. READ
FRANKLIN SOGANDARES-BERNAL
HELEN E. STALLARD
GEORGE L. STEWART
L. T. THREADGOLD
GARY L. UGLEM

PUBLISHED BY
WILLIAM MARSH RICE UNIVERSITY
HOUSTON, TEXAS

Vol. 62, No. 4

Fall 1976

US-ISSN-0035-4996
US-ISBN-0-89263-230-5

CLARK PHARES READ
1921-1973

Professor Clark Phares Read died unexpectedly in Houston, Texas, on December 23, 1973. He had been born the son of Helen and Clark Phares Read in Fort Worth, Texas, on February 4, 1921.

Clark Read served in the U. S. Navy and Marine Corps during 1940-1945. His university career was initiated at Tulane University in 1943, where he was assigned as a student in the old U. S. Navy V-12 program until 1945. He then attended the University of Texas Medical School until his resignation in 1946. Read's classmates and professors were shocked to hear of his resignation from medical school; but aided and abetted by Dean Chauncey Leake he opted for a research career, a fortuitous event for the field of parasitology. He had been introduced to research at Tulane by Professor Edward S. Hathaway, an exacting taskmaster to those of us who came to know him well. For a time it appeared as if the young scientist's choice of research career might not pan out. He had applied to work with Ernest Carroll Faust at Tulane, but was not accepted because Professor Faust already had too many graduate students. Read then applied to work under the direction of Asa C. Chandler at Rice Institute. Professor Chandler was very demanding of his students, and his new pupil developed rapidly under his direction. As Read himself admitted in 1959, he had found himself thriving under pressure. Since Clark had not formally completed a bachelor's degree, Professor Chandler required him to write and defend two master's theses. By 1950 he had earned the Ph.D. and left Rice to join the faculty of the University of California at Los Angeles, where he held the rank of Instructor of Zoology. Read earned the title of Assistant Professor in 1952. At U.C.L.A. he came under the influence of an outstanding scientist and formidable individual, Gordon Ball. Clark fondly remembered the association in 1959 by saying,

> In my eyes Dr. Ball is a real giant of personal and scientific integrity. When I joined the faculty at U.C.L.A. as an instructor in 1950, Dr. Ball was Professor of Zoology. A lesser man than Gordon Ball would have expected a novice to assume a junior role in the scheme of things. However, Gordon will give you as much rope as you want. When you become entangled he will cut the rope and hand it back to you. We shared many experiences in

teaching and departmental duties and I profited immeasurably from the intangible assets of Professor Ball's personality.

In 1954 Read joined the faculty of the Johns Hopkins University in Pathobiology, which was headed by Frederik Bang, an individual for whom he came to have great affection. In 1959, already the author of at least sixty papers primarily dealing with biochemistry and physiology, and the mentor of sixteen Ph.D. candidates, he joined the then Rice Institute faculty as professor of biology—a post he held until his death.

The same year he arrived at Rice, Read was designated the first recipient of the Henry Baldwin Ward Medal of the American Society of Parasitologists. This medal is presented to young parasitologists who have demonstrated excellence in scholarly activities and original research.

National and international recognition of his work was followed by a Guggenheim fellowship, in 1960, to Cambridge University, where he was designated Research Fellow of Christ's College, and where he worked at the Molteno Institute. From 1961 to 1966 he was Chairman of the Zoology Department at the Marine Biological Laboratory at Woods Hole, Massachusetts, a laboratory to which he moved every possible summer thereafter to conduct research. Later he was appointed a Trustee (1966-1974) of the Marine Biological Laboratory. From 1961 to 1966 Read served as Chairman of the Study Section in Tropical Medicine and Parasitology, National Institute of Allergy and Infectious Diseases of the National Institutes of Health. In 1963 he was Carnegie Professor at the University of Hawaii. He also held a Career Professorship from the National Institutes of Health in 1966.

Professor Read served as Council Member at Large of the American Society of Parasitologists from 1960 to 1963, and served on the editorial boards of the *American Journal of Hygiene, Experimental Parasitology, Journal of Comparative Biochemistry and Physiology, Parasitology,* and *Journal of Parasitology.* He was also a member of the American Association of University Professors, the Society of General Physiologists, the American Society of Zoologists, the American Society of Parasitologists, the American Society of Tropical Medicine and Hygiene, the Society of Protozoologists, the American Physiological Society, the Society of the Sigma Xi, and the Southwestern Association of Parasitologists; he was a Fellow of the Royal Society of Tropical Medicine.

Elsewhere in this memorial volume, George Stewart has reviewed the scientific contributions of Professor Read in considerable detail. It may be said, however, that Clark Read's greatest contribution to the field of parasitology, besides the training of many excellent students who now hold positions of leadership in the field, was that of bringing methodologies and concepts from other fields into parasitology. He was one person, among a

handful of others, instrumental in bringing the physiological bases of
parasitism to a level of sophistication acceptable to biochemists and physiologists alike. It was no easy task. For a time Read was greatly distressed by
the grip of the old "autocrats" in which the American Society of Parasitologists found itself, and by the unwillingness of the traditional classicists
to apply new concepts and techniques from other scientific fields to problems
in parasitology. For a period of several years he failed to appear at national
parasitology meetings. I once chided him about his absence from these
meetings, arguing that if we did not work for change, it could not be expected
to come about. He replied that he had tried, but the situation was simply
hopeless. Clark Read understood and appreciated the traditional parasitology of the times and he published several excellent papers in systematics.
I mention this because that difficult era was dominated by parasitologists
who were traditional taxonomists. Read was not disdainful of taxonomy;
he simply recognized that insufficient progress was being made and that
more sophisticated techniques needed to be applied to problems in parasitology to keep our field abreast of other rapidly moving disciplines in biology.
Read's decision to end his self-imposed exile from national meetings came
a few years before his death, and was greatly influenced by the constant
encouragement he received from his many former students, colleagues, and
friends in the Southwestern Association of Parasitologists (SWAP). After the
second meeting of SWAP held at Tulane University, he conveyed to me his
great satisfaction at the high quality of the papers presented. He never failed
to attend a SWAP meeting after the formation of the group, and was its
president at the time of his death. His renewed interest in the national society
was, in my opinion, sparked by the fact that he was no longer a lone voice
far ahead of his time. Read constantly fought to improve the quality of work
in our field, and in a 1963 publication he made the following appeal to
parasitologists:

> Because of the general subject of this monograph, an additional liberty will be taken in
> making a few remarks concerning the training of persons for research in the biochemical
> and biophysical aspects of parasitism. It has become apparent that it is extremely difficult,
> if not impossible, to furnish adequate training for research in biochemistry, as well as a
> thorough grounding in the morphology and biology of parasitic organisms. The training
> of students for a research career implies preparation in depth. Competent work in
> morphology requires a thorough grounding in morphology with sufficient biochemistry
> to appreciate progress in biology. Competent work in the biochemistry of parasitism
> requires training in chemistry, mathematics, and biochemistry with a sufficient back-
> ground in parasitology to engender appreciation of the chemical problems of greatest
> biological importance. Several years ago, the world's senior citizen in the biochemistry
> of parasitism, Theodor von Brand, warned of the dangers of jumping on band wagons.
> We should take this warning seriously. Not all of us have taken it seriously. Our field
> has suffered from a spate of superficial studies in biochemistry and physiology. There is no

longer an excuse for the publication of papers in our field which do not meet the requirements of quality and rigorous scientific interpretation expected of papers published in such journals as *The Journal of Biological Chemistry, The Biochemical Journal, The American Journal of Physiology, The Journal of General Physiology*, and perhaps a half dozen other such journals. We must demand quality regardless of questions of quantity. An investigator in our field who publishes one good paper a year must stand above those who choose to publish three or more inferior papers in a similar period. We have been beset in the western world during the past 15 years with the problem of keeping up with the literature. I would contend that this problem would be considerably simplified if we placed a greater emphasis on quality and depreciated the emphasis on quantity. Growth of the field with which we are concerned requires such an emphasis. If we fail to impose this emphasis we shall follow the path of certain other underdeveloped areas of biology. We shall fail to attract competence and talent. We are at a very important and crucial point in the development of the experimental study of parasitism. We have inadequate communication with the well-developed areas of medical microbiology and plant pathology, as well as the highly sophisticated and complex areas of biochemistry, biophysics, and general physiology. Ours is a grave responsibility. Those of us who are responsible for the training and guidance of others must accept this responsibility and work at discharging it in a competent fashion. If we fail to do so, others will do it for us. We will be abdicating our responsibilities and, in more crass terms, our meager claims to fame. There are some persons who have furnished us examples of rigor and quality in this field. I would point to the beautiful and sophisticated researches of von Brand, Fairbairn, Bueding, Saz, Campbell, Kmetec, Entner, Hutner, Nathan, and others in animal parasitology. We could learn much from the researches of Beck, Schneider, Braun, Bloch, Charniaux-Cotton, Sonneborn, Garber, Lewis, and others in those areas which superficially appear to be peripheral to our own. Ours is an exciting field with fantastic possibilities for good minds. We are dealing with some of the most complex systems in biology. We are concerned with the interactions of two steady states. We are concerned with the emergent properties, not readily predictable, of such interactions. We must enlist the talents of physicists, of mathematicians and most importantly, of competent biologists to attack these problems and to train our students. A course of action requires a renewed dedication to our interests.

Many scientists are content to work quietly in their laboratories. The same could never be said of Clark Read. His deeply ingrained sense of justice and constitutional rights of citizens led him to serve as Secretary and Chairman (1967-1970) of the Houston Chapter of the American Civil Liberties Union, and Vice President of the Texas Civil Liberties Union (1970-1972). From 1968 to 1972, he served as Chairman of the Free Lee Otis Johnson Committee. He also served as a member of the boards of the Martin Luther King Foundation and the Houston Council of Human Relations, and as a sponsor for the Fund of the Republic and for the United Farm Workers Seminar. His concern for the education of disadvantaged citizens led him to serve as a member of the Advisory Board for the Mexican-American Educational Committee and of the Desegregation Advisory Committee of the Houston Independent School District.

His other activities were varied, including service as the principal scientific advisor to Roberto Rossellini on a ten-hour television film titled "Science."

He also served as principal scientific advisor on a United Nations film titled "Populations—What and Where."

How did all of these activities, honors, and responsibilities affect Read? He was a warm individual completely lacking in pretension. Until his death he retained the ability to laugh at himself. He was a generous person to all who were fortunate enough to know him. In later years he was an outstanding spokesman for academic freedom and the rights of young people everywhere. He loved life and a challenge.

Clark dearly loved his wife, Lee, and his children, Johanna Read Tobias, Victoria Read Miller, and Thomas Jefferson Read. Shortly before his death, he wrote me a long letter to tell me of his wife's most recent success of which he was very proud. Lee, the love of his life, had just received a law degree. Read admired her greatly and he wrote, "she is one of the most intelligent persons I have ever known."

The American Civil Liberties Union and Rice University have both established memorial funds in the memory of this Texas giant. There will never be another person quite like Clark Read, but upcoming young people everywhere can take a lesson on how to mold their lives from a man who remains a living memorial to distinction.

FRANKLIN SOGANDARES-BERNAL
DEPARTMENT OF BIOLOGY
SOUTHERN METHODIST UNIVERSITY

RICE UNIVERSITY STUDIES

Vol. 62, No. 4 Fall 1976

STUDIES IN PARASITOLOGY

In memory of
CLARK P. READ

CLARK PHARES READ
by Franklin Sogandares - Bernal v

THE CONTRIBUTIONS OF CLARK P. READ
ON THE ECOLOGY OF THE VERTEBRATE
GUT AND ITS PARASITES
*by George L. Stewart, Richard D. Lumsden, and
Frank M. Fisher, Jr.* 1

A UNIQUE TEGUMENTARY CELL TYPE AND
UNICELLULAR GLANDS ASSOCIATED WITH THE
SCOLEX OF *EUBOTHRIUM CRASSUM*
(CESTODA: PSEUDOPHYLLIDEA)
by C. Arme and L. T. Threadgold 21

AMINO ACID POOLS OF *SCHISTOSOMA MANSONI*
AND MOUSE HEPATIC PORTAL SERUM
by Harold L. Asch 35

ACANTHOCEPHALAN DEVELOPMENT: MORPHOGENESIS
OF LARVAL *MONILIFORMIS DUBIUS*
by J. E. Byram and Kay W. Byram 43

THE BIOLOGICAL SIGNIFICANCE OF THE TEGUMENT
IN DIGENETIC TREMATODES
by K. E. Dixon 69

BIOCHEMICAL AND CYTOCHEMICAL STUDIES OF
ALKALINE PHOSPHATASE ACTIVITY IN
SCHISTOSOMA MANSONI
by Sue Carlisle Ernst 81

PROLINE IN FASCIOLIASIS: II. CHARACTERISTICS OF
PARTIALLY PURIFIED ORNITHINE-δ-TRANSAMINASE
FROM *FASCIOLA*
by Jeanette C. Ertel and Hadar Isseroff 97

ULTRASTRUCTURAL CYTOCHEMISTRY OF THE
TEGUMENTAL SURFACE MEMBRANE OF
PARAGONIMUS KELLICOTTI
  by Francis M. Gress and Richard D. Lumsden 111

THE EFFECT OF CORTISONE ON THE SURVIVAL OF
HYMENOLEPIS DIMINUTA IN MICE
  by C. A. Hopkins and Helen E. Stallard 145

DENSITY DISTRIBUTION OF DNA FROM PARASITIC
HELMINTHS WITH SPECIAL REFERENCE TO
ASCARIS LUMBRICOIDES
  by Araxie Kilejian and Austin J. MacInnis 161

ULTRASTRUCTURAL CHANGES IN THE INFECTIVE
LARVAE OF *NIPPOSTRONGYLUS BRASILIENSIS* IN
THE SKIN OF IMMUNE MICE
  by D. L. Lee 175

SPECIFICITY OF AMINO ACID TRANSPORT IN THE
TAPEWORM *HYMENOLEPIS DIMINUTA* AND
ITS RAT HOST
  by A. J. MacInnis, D. J. Graff, A. Kilejian, and C. P. Read 183

EFFECTS OF CARBON DIOXIDE ON GLUCOSE
INCORPORATION IN FLATWORMS
  by James S. McDaniel, Austin J. MacInnis, and Clark P. Read 205

STUDIES ON BIOCHEMICAL PATHOLOGY IN
TRICHINOSIS. II. CHANGES IN LIVER AND
MUSCLE GLYCOGEN AND SOME BLOOD
CHEMICAL PARAMETERS IN MICE
  by George L. Stewart 211

ABSORPTION KINETICS OF SOME PURINES,
PYRIMIDINES, AND NUCLEOSIDES IN
TAENIA CRASSICEPS LARVAE
  by Gary L. Uglem and Michael G. Levy 225

THE CONTRIBUTIONS OF CLARK P. READ
ON THE ECOLOGY OF THE
VERTEBRATE GUT AND ITS PARASITES

by George L. Stewart, Richard D. Lumsden, and
Frank M. Fisher, Jr.

Clark Read's scientific contributions had great impact on the discipline of parasitology and on biology as a whole. It was Read, more than any other worker, who articulated the concept of the host-parasite interface and defined many of the intricate molecular interactions between symbiotes and their hosts. As Justus Mueller, editor of the *Journal of Parasitology*, recently stated, "Clark Read was Parasitology's ambassador to the fields of physiology, biochemistry and molecular biology."

Some of Read's greatest contributions dealt with physiological interactions between intestinal-parasitic helminths and the vertebrate gut, and it is to these works that the following paragraphs are addressed. In 1950, while a graduate student at the Rice Institute, Clark Read completed a monograph entitled *The Vertebrate Small Intestine as an Environment for Parasitic Helminths,* in which he dealt with the parameters of intestinal physiology of greatest importance to the parasitologist. In this publication Read devoted particular attention to the dynamics of the intestinal environment. He considered contributions made to the gut contents by the stomach, liver, pancreas, and intestinal mucosa, with special reference to the importance of these organs in conditioning the gut lumen as an environment for parasites. He presented strong support for the existence of an exocrino-enteric circulation involving the flow of materials, with the notable exception of carbohydrates, from the blood and from other organs into the lumen of the vertebrate intestine, and subsequent resorption of these materials by the intestine. Read concluded that this flow of organic compounds from the tissues, much of which would be resorbed in areas of the gut distal to the point of secretion, would be available to lumen-dwelling parasites. In addition, Read emphasized the existence of lateral variation in the physico-chemical conditions of the

George Stewart is Lecturer in Biology at Rice University. Richard Lumsden is Professor of Biology and Dean of Graduate Studies at Tulane University. Frank Fisher is Professor of Biology at Rice University.

intestine, which resulted in a luminal environment with one set of parameters, and a paramucosal environment possessing a different set of physico-chemical conditions. He defined the paramucosal-lumen as "that portion of the lumen of the intestine which is immediately adjacent to the mucosa," and reviewed research that demonstrated differences in the pH, osmotic pressure, oxygen tension, and concentrations of organic and inorganic substances between this area and the center of the intestinal lumen. He also described the paramucosal-lumen as an area of the intestine that may have a close physiological resemblance to the intracellular spaces of the host.

Later, from the work of others and from the results of studies conducted in his laboratory, Read (1970) pointed out that, independent of the quality of food ingested, there is great stability in the molar ratios of free amino acids found in the lumen of the small intestine of the dog, the rat, and the dogfish (Nasset et al., 1955; Nasset and Ju, 1961; Nasset, 1962; Read, Simmons, and Rothman, 1960; Simmons, unpublished). Read (1970) upheld the contention that most important among the sources of endogenous nitrogen contributing to stability of the molar ratios of amino acids in the gut of vertebrates are: 1) digestive enzymes and other secreted proteins found in the saliva, gastric juice, pancreatic juice, intestinal juice, and bile; 2) sloughed off cells of the intestinal epithelium; and 3) the bidirectional flux of free amino acids across the mucosa. In addition, it was noted that fatty acids of the gut were diluted by lipids from exogenous sources, although homeostasis was not as precise as that seen with amino acids (Ginger and Fairbairn, 1966; Kilejian et al., 1968). Maintenance of stable amino acid and fatty acid molar ratios in the intestinal lumen, Read (1970) stated, was an extremely important characteristic of the environment of gut parasites. As was demonstrated by Read, Rothman, and Simmons (1963) for the tapeworm *Hymenolepis diminuta*, competition between different molecular species in mediated absorption (active transport) is a function of molar ratios and not of absolute quantities. Read stated that

> the membrane transport systems of the parasite are exposed to relatively constant molar ratios, and the molar ratios of amino acids absorbed must remain relatively constant in a given host. . . . the parasite, at least in the case of tapeworms, seems to have no mechanisms for establishing or maintaining relative constancy in amino acid ratios relative to the external surface. We are led to the almost inescapable conclusion that parasitism by *Hymenolepis* is not only concerned with the food obtained from the host, but with parasitism of those mechanisms of the host concerned with regulating the molar ratios of amino acids. It is suggested that the basis of this parasitism is considerably more subtle than the obtaining of simple chemical compounds to satisfy nutritional requirements. Unquestionably, nutritional requirements must be and are met, but of similar significance will be dependence on maintenance by the host of a mixture of nutrients compatible with the coupling of absorption and synthetic mechanisms. It may be said that competition between amino acids for transport into cells is clearly of importance for worm metabolism, even though the actual amino acid requirements are not known. Although a given amino acid might not be required by the worm, the presence of this amino acid may regulate, by competition, the rate at which a required amino acid is taken up from the medium.

Consideration of the relationship of the membrane transport systems of tapeworms to the maintained steady state amino acid ratios of the environment leads to the conclusion that these worms parasitize a physiological control mechanism of the host. Many animal parasites, almost certainly tapeworms, may have certain regulatory capacities lacking or rudimentary. This implies a functional integration with the host which may prove to be of greater consequence than such features as novel, absolute nutritional requirements. It would represent marked integration with regulatory mechanisms of the environment, rather than independence from the environment, for a maintenance of a steady state in the parasite. (1970:190)

Clark Read had long contended (1950:77) that "In order to understand the host-parasite relationships of intestinal helminths we must separately investigate the physiology of the host and of the parasite. . . . Following such study a resynthesis of host and helminth physiology will reveal entirely new concepts relative to intestinal parasitism." Along these lines Dr. Read devoted much of his scientific energy, from 1950 to the time of his death, to study of the physiology of helminth parasites of the vertebrate gut.

Between 1952 and 1959 Read published several papers dealing with the enzymology, carbohydrate metabolism, and biology of tapeworms. After assuming his first faculty position at U.C.L.A., he used biochemical techniques to demonstrate the presence of cytochrome oxidase and succinic dehydrogenase in the rat tapeworm, *Hymenolepis diminuta*. This was the first report of the former enzyme from a cestode (Read, 1952). An investigation of anaerobic dehydrogenases in the same tapeworm (Read, 1953) demonstrated pyridine-nucleotide-linked enzymes (catalyzing the oxidation of a number of different organic acids) and fumarase and cytochrome-linked α-glycerophosphoric dehydrogenases.

Upon accepting a position at the School of Public Health at Johns Hopkins University, Clark Read began work on the carbohydrate metabolism of helminth parasites of the vertebrate gut. He (1956) showed that in *H. diminuta* aerobic respiration was stimulated or supported by glucose, succinate, malate, glycerophosphate, and malonate. The end-products of anaerobic metabolism in *Hymenolepis* were delineated, and Read showed that this worm carried out glycolysis independent of the glucose concentration in the medium. In reference to the latter observation he stated, "This indicates that the worm is capable of deriving benefit from glucose at very low concentrations." As will be seen, it was through his studies of the "active transport" of nutrients by helminth parasites that Read made some of his greatest contributions to biology.

To determine the function of carbohydrate metabolism in the microecology of helminth parasites of the vertebrate gut, Read began a series of studies entitled "The role of carbohydrates in the biology of cestodes" (Read and Rothman, 1957a, b, and c). The size of the tapeworm *H. diminuta* was affected by the quality of carbohydrate ingested by the host. Significant differences in the volume of gravid segments were noted in worms from hosts allowed to consume different carbohydrates. Morphological changes induced by specific carbohydrates were accompanied by changes in the reproductive rates of worms. *Hymenolepis* from hosts on diets containing only starch, of all the carbohydrates tested, demonstrated the best size maintenance and rates of reproduction. Less effective in supporting worm growth and reproduction were host diets in which glucose, dextrins-maltose, or sucrose was the sole dietary carbohydrate. Fructose alone depressed worm reproduction to a level equal to that of worms from hosts receiving no carbohydrate at all in their diets. Eggs produced by worms from hosts on a sucrose diet were abnormal in shape and reduced in size. The reproductive rates and growth of worms from hosts receiving suboptimal amounts of dietary starch, plus fructose, were greater than those of worms from hosts on a similar starch diet without fructose. Read (1959) suggested that fructose may interfere with absorption by the host gut of the products of starch hydrolysis. The quality of carbohydrate in the host's diet also affected the growth of *Hymenolepis citelli* and *Hymenolepis nana*. As in the case of *H. diminuta*, sucrose did not support the growth of these worms as well as starch (Read, Schiller, and Phifer, 1958). Reduction in the size and number of *Lacistorhynchus tenuis* occurred in dogfish starved for seven days. When the host was given carbohydrate orally during the starvation period, however, the size and number of this tapeworm were normal (Read, 1957). Read and Phifer (1959) tested the hypothesis that competition for usable carbohydrates by tapeworms is the factor responsible for limiting the size of individual worms in infections of varying intensities. They found that under conditions of crowding, in hosts given a worm-limiting amount of starch or sucrose, the weight of individual *H. diminuta* decreased in the presence of increasing numbers of worms. In this same study animals were infected with one worm of each of two species of tapeworm (*H. diminuta* and *H. citelli*). *Hymenolepis citelli* from hosts on low-starch diets showed an equal reduction in size whether or not *H. diminuta* was present. *H. diminuta*, on the other hand, became smaller in the presence of *H. citelli* than in its absence (Read, 1951). Read and Rothman (1958a) observed a marked reduction in weight of the acanthocephalan *Moniliformis dubius* from rats fed a low-carbohydrate diet. The polysaccharide content of worms was dramatically affected by the quality of carbohydrate included in the host diet. Diurnal fluctuations in the polysaccharide content of *Moniliformis* were correlated with the feeding habits of the host.

Although some six or seven sugars were examined, *Hymenolepis diminuta* (Read, 1956), *H. nana*, *H. citelli*, *Mesocestoides latus* (Read and Rothman, 1958b), *Calliobothrium verticillatum*, and *L. tenuis* (Read, 1957) metabolized only glucose and galactose to a significant extent. On the other hand, *Cittotaenia* sp. (Read and Rothman, 1958b) utilized maltose as well as sucrose.

In the final paper of this series of publications dealing with carbohydrate metabolism in cestodes, Read (1959) presented the following conclusions and hypotheses: 1) carbohydrates in the gut contents of the host are used by tapeworms to fulfill carbohydrate requirements for their growth and reproduction; 2) there are strict limitations in the quality of carbohydrates that tapeworms can utilize to support growth and reproduction; 3) the rate of growth and reproduction and the size attained by worms is dependent on the type of carbohydrate eaten by the host; 4) competition for available carbohydrate in the host gut may be the basis for the crowding effect seen between worms of the same and of different species; 5) of the three carbohydrates tested (glucose, sucrose, and starch), starch was best in supporting the growth and reproduction of tapeworms. Read concluded that since glucose appeared to be rapidly absorbed by the host in the upper part of the small intestine, and since the enzyme sucrase was primarily located in the same area of the vertebrate gut, the quantity of carbohydrate usable by the worm in hosts on diets including only glucose or sucrose was limited. On the other hand, the more complicated series of hydrolytic events involved in the degradation of starch to glucose would allow passage of usable sugar to the area of the gut occupied by tapeworms and would present worms with usable sugar for longer periods of time; 6) the imposition of the dynamics of vertebrate gut physiology on the specific carbohydrate requirements of tapeworms is of importance in cestode distribution, host age, or sex-linked resistance and speciation in tapeworms. The feeding habits of the host are apparently of importance to establishment of a cestode in a particular host as well as to the completion of the tapeworm life cycle; and 7) competition between host and tapeworm for carbohydrates appears to be negligible, since the parasites are located below the region in the host gut in which greatest absorption of carbohydrates occurs.

Additional studies on the influence of host feeding habits on the biology of tapeworms were carried out in Read's laboratory. *Hymenolepis diminuta* underwent a circadian longitudinal migration in the small intestine of the rat, which was related to the feeding of the host (Read and Kilejian, 1969). Further observations on this phenomenon (Chappell, Arai, et al., 1970) revealed the following three migration phases on the basis of worm age: 1) five- to seven-day-old worms migrated from the middle of the small intestine to its anterior end during host fasting; 2) seven- to eight-day-old worms showed no migration; 3) nine- to fourteen-day-old worms migrated to the posterior end of the host small intestine during host fasting and anteriorly during host feeding. A relationship between the migratory behavior of worms, worm growth, and the "crowding effect" was observed. Read postulated that migratory phase 3 and the crowding effect resulted from intraspecific, and perhaps interspecific, competition for nutrients, probably glucose. His hypothesis was supported by the fact that glycogenesis was minimal in worms older than seven days during host fasting and was greatest during host feeding. In addition, glycogen levels in young worms fluctuated regularly, whereas glycogen levels in older worms changed diurnally. This difference implied that intestinal glucose concentrations are not limiting to growth in *H. diminuta* until the worms reach a certain mass, at which time competition for glucose influences the positioning of worms in the gut. In smaller, younger worms such competition may not exist, and worm positioning is determined on the basis of the best location in relation to glucose concentrations in the gut as dictated by host feeding habits.

Read, Rothman, and Simmons (1963) stated, "The physicochemical relationships between a host and a parasite will involve a region in space and time which may be termed the host-parasite interface. The host-parasite interface will be intimately involved in determining the nature and extent of integration, and thus the outcome of the relationship, since it represents the region of chemical juxtaposition of regulatory mechanisms of both host and parasite. In the latter regard, two distinct aspects of the interface have regulatory significance: the physicochemical characteristics of the surrounding host fluids, together with mechanisms regulating their composition; and the functional characteristics of surfaces or surface membranes of the parasites themselves. The latter includes the bounding membrane and associated organelles in acellular forms." Parasitism is usually defined as a nutritional relationship. The process by which parasites obtain food should therefore be considered of extreme importance in understanding relationships between parasites and their environment. In order for a material to serve as a nutrient it must be available for metabolism. Since tapeworms possess no digestive tract, it was assumed that nutrients entered the body through the outside surface. The absorptive nature of the external surface of tapeworms had been demonstrated in electron microscopic studies (Rothman, 1963; Lumsden, 1965). Lumsden (ibid.) concluded that the microvilli found on the surface of tapeworms were quite similar to those seen on the surfaces of cells with known absorptive functions, such as the epithelial cells of the vertebrate intestine. Furthermore, the tapeworm tegument possessed physiological activity (see review by Read, 1966) found associated with other absorptive surfaces.

After 1959 much of the research carried out in Read's laboratory aimed at understanding animal parasitism with respect to the chemical exchanges between hosts and parasites. As mentioned earlier, Read considered the host-parasite interface to be a highly significant component of these chemical exchanges. An understanding of host-parasite nutritional relationships on the molecular level would require knowledge of the processes involved in the movement of organic compounds across the parasite surface (Read, Rothman, and Simmons, 1963). Most of the work done on these processes in animal parasites was conducted in Clark Read's laboratory, or in the laboratories of scientists who had studied under him. Read pursued the following objectives in the studies discussed below: 1) a detailed analysis of membrane transport systems in helminth parasites of the vertebrate gut; 2) study and analysis of the interactions of extrinsic host enzymes with the surface of gut parasites; and 3) study and characterization of intrinsic-bound parasite surface enzymes, with special reference to interaction of such enzymes with membrane transport systems.

Read, Rothman, and Simmons defined "active transport" as a

term applied to processes exhibiting characteristics which are quite different from those of simple diffusion. The most significant aspect of the process is the movement of a substance across a membrane against a chemical concentration difference—i.e., there is a net movement of solute not attributable to its kinetic energy. Such uphill transport implies the participation of forces other than those of diffusion, and at the present time is the only certain criterion of active transport. Usually, however, the process is characterized by stereospecificity involving competitive inhibitions by chemically similar compounds and inhibitions by poisons of energy metabolism. Rather than being a linear function of concentration difference, the rate of movement of a substance being actively transported most frequently follows saturation kinetics. (1963:155)

Read, Simmons, Campbell, and Rothman (1960) demonstrated the presence of a mediated system of entry (active transport) for amino acids in the gut of the dogfish and the tapeworm *Calliobothrium verticillatum*. Definite differences between the amino acid transport systems of the two were shown; Read and Simmons (1962), however, revealed that competitive inhibition of the uptake of a single amino acid by a mixture of amino acids was similar in kinetic characteristics to that seen in inhibition by a single amino acid. Uptake of some amino acids by *H. diminuta* (Kilejian, 1966; Harris and Read, 1968), *C. verticillatum* (Read et al., unpublished), and the dogfish gut (Read, 1967) were inhibited by previous accumulation of sugars which, in the case of the dogfish intestine, were actively absorbed by the mucosa. Similar to a number of other tissues and organs (Crane, 1965), the dogfish intestine (Read, 1967) possessed sodium-dependent amino acid and sugar transport systems. Potassium interfered with the sugar, but not the amino acid, entry system. Read, Rothman, and Simmons (1963) demonstrated active absorption of amino acids by *H. diminuta* and showed conclusively that various amino acids competed with each other in the process of entry. Study of competition between amino acids for entry into worms, and use of several metabolic poisons, led to proposal of four qualitatively different loci for the active absorption of amino acids by *H. diminuta*. Work with worms of different ages and from different hosts led to the conclusion that important phenotypic changes in the active entry systems for amino acids may occur from exposure to environments provided by different hosts, or as a function of differences in the physiology of worms of different ages. On the basis of these studies, the researchers hypothesized (ibid.) that observed alterations in the active transport systems in *H. diminuta* may be a result of changes in the relative numbers of qualitatively different transport loci for amino acids. Further characterization of the systems for active transport of amino acids in *H. diminuta* was carried out by Harris and Read (1968), Laws and Read (1969), Woodward and Read (1969), and Pappas, Uglem and Read (1974).

Harris and Read (1969) found that *H. diminuta* incorporated amino acids from the surrounding medium into protein. The sharp decrease in the polysaccharide reserves of starved worms, shown by Read (1956), was accompanied by a decrease in the incorporation of amino acids into the protein of *H. diminuta* (Harris and Read, 1969). Glucose stimulated the incorporation of lysine into proteins of starved *H. diminuta* and of valine into proteins of starved *C. verticillatum*, but inhibited incorporation of valine into the proteins of unstarved *C. verticillatum* (Fisher and Read, 1971). Glucose also inhibited incorporation of lysine into proteins of unstarved *H. diminuta* (Harris and Read, 1969). The effects of available carbohydrate on protein synthesis in *H. diminuta* supported the previous finding of Read (1956) that carbohydrate was required for growth and reproduction in that parasite. Harris and Read concluded that *H. diminuta* possesses a "catabolic carbohydrate metabolism driving an anabolic protein and fat metabolism" (1969:654).

Arme and Read (1969) investigated the host-parasite relationship between the rat and *H. diminuta*, concentrating on the distribution of the non-metabolized amino acids in the host and the parasite, and on bidirectional fluxes of these amino acids *in vivo* and *in vitro*. Following injection into rats, the non-metabolizable amino acid cycloleucine quickly reached a "steady-state distribution" in the tissues of the host and of the parasite. The concentration of cycloleucine was lower in worms than in host serum. High rates of exchange between the host and the parasite were comparable to those seen between the organs and body fluids of the rat. The investigators concluded from these findings that tapeworms parasitizing the vertebrate gut fulfill their amino acid requirements from endogenous host sources, and further, that "the rates of flow of cycloleucine into the gut lumen, and the apparent concentration of this amino acid rapidly attained, together form a powerful argument that the worm lives in an extracellular space which allows access to the amino acid pool of the host body." In addition, they reported that: 1) there was a rapid flow of non-protein amino acids (cycloleucine and α-amino-isobutyric acid) into the lumen of the rat intestine, and the concentrations rapidly reached equilibrium with those of the extraintestinal body fluids, and 2) administration of a single amino acid of sufficient quantity would produce a gross imbalance in the amino acid composition of the intestinal contents resulting in a rapid influx of endogenous free amino acids. On the basis of these findings, and of the results of studies by Simmons,

Read, and Rothman (1960), Hopkins and Callow (1965), and Nasset and Ju (1961), it was postulated that "the flow of endogenous free amino acids, rather than the digestion of endogenous protein, may be of greater importance in regulating the relative quantities of free amino acids in the gut lumen." In addition, the active influx and efflux of cycloleucine, and interactions of other amino acids with the absorption and outflow mechanisms for this amino acid in the gut of the rat host and in *H. diminuta*, were determined and compared. In contrast to findings with *H. diminuta*, lysine and proline inhibited the uptake of cycloleucine and stimulated its efflux from rat gut.

The active absorption of glucose by intestinal helminths was demonstrated for *H. diminuta* (Phifer, 1960) and *C. verticillatum* (Fisher and Read, 1971). Studies on the role of cations in glucose uptake by *H. diminuta* (Read, Stewart, and Pappas, 1974) and *C. verticillatum* (Fisher and Read, 1971) revealed that the active entry system for this sugar in both worms was sodium sensitive. Read (1961) found that the uptake of glucose by *H. diminuta* was competitively inhibited by galactose and some other monosaccharides. These findings suggested that the glucose transport system in *H. diminuta* was similar to that of the vertebrate intestine (reviewed by Crane, 1960). Fisher and Read (1971) found that *C. verticillatum* took up glucose and galactose but did not transport or metabolize mannose or fructose. The gas phase had no effect on the absorption or accumulation of glucose by this tapeworm, and the optimal pH for the transport of glucose (pH 8.9) was the same as that found in the part of the dogfish gut inhabited by *Calliobothrium*.

Arme and Read (1968) presented evidence for an active transport system specific for short chain fatty acids (less than nine carbon atoms in the hydrocarbon chain) in *H. diminuta*, and Chappell, Arme, and Read (1969) showed this tapeworm to possess a mediated system of entry specific for long chain fatty acids (greater than twelve carbons in the hydrocarbon chain). In the latter study, the uptake of ^{14}C-palmitate was markedly stimulated by the presence of a number of long chain fatty acids. Stimulation occurred only when the concentration of the effector molecules (stimulator) was present at the same concentration as palmitate. The researchers observed a second type of stimulation of palmitate transport, by laurate, which stimulated the uptake of palmitate regardless of the concentration of laurate. Laurate stimulation was considered to be a result of the unique ability of this compound to increase the solubility of fatty acids (Fieser and Fieser, 1959).

The transport of purines and pyrimidines into *H. diminuta* was first investigated by MacInnis, Fisher, and Read (1965), who demonstrated mediated processes for the uptake of several of these compounds and showed inhibitory and stimulatory interactions between structurally analogous molecular species. Pappas, Uglem, and Read (1973b) further characterized the purine-pyrimidine transport systems in *Hymenolepis* and proposed a

3-locus model for the transport of purines and pyrimidines based on inhibitory and stimulatory interactions between the various compounds tested. McCracken et al. (1975) have found that the mediated absorption of certain pyrimidine nucleosides by *H. diminuta* is Na^+ dependent.

Evidence that *H. diminuta* requires water-soluble (B) vitamins for normal growth and development was presented by Platzer and Roberts (1969, 1970). Pappas and Read (1972a, b) investigated the mechanisms for entry of thiamine (B_1), riboflavin (B_2), and pyridoxine into *H. diminuta*. These workers demonstrated the existence of both mediated and non-mediated systems for entry of thiamine into this tapeworm. Riboflavin was absorbed by a specific, active process, and the mediated uptake of the vitamin was inhibited by a number of compounds structurally similar to riboflavin. The absorption of pyridoxine occurred by diffusion. Pappas and Read (1972a, b) pointed out some similarities between vitamin absorption by *H. diminuta* and the mammalian gut. *H. diminuta* transports thiamine and riboflavin and obtains pyridoxine by diffusion, whereas the mammalian intestine actively absorbs only thiamine (Matthews, 1967) and takes up pyridoxine by a non-mediated process. Evidence was presented in support of the hypothesis that vitamins are supplied to worms *in vivo* by the exocrino-enteric circulation of the host.

Read (1970) pointed out that the vertebrate mucosa possesses a brush border facing the gut lumen which has been widely considered to be involved primarily in the absorption of nutrients by the host. Recent evidence, however, (Nachlas et al., 1960; Miller and Crane, 1961; Eichholz, 1967; Eichholz and Crane, 1965; Johnson, 1967) has demonstrated that this surface is also endowed with a capacity for digestion of proteins and a wide variety of sugars. In addition, Read mentioned that a charged mucopolysaccharide layer (glycocalyx) on the surface of the mucosa cells (Ito, 1969) was capable of binding proteins (Bell, 1962) and conferred additional digestive capacity on these cells. This mechanism of potentiation of digestive capacity following adsorption of proteins to the surface of mucosa cells had been referred to by Ugelov (1965) as "contact" digestion. Read (1970) emphasized the potential for a close functional relationship between the absorptive and digestive activities present on the surface of the cells of the mucosa, and he mentioned the work of Crane (1967), which showed that a favorable spatial arrangement of absorptive and digestive components of the membrane would result in a "kinetic advantage" for the absorption of the products of sugar hydrolysis. As has been pointed out earlier, the tegument of *H. diminuta* shows functional and morphological similarities to the mucosa cells of the vertebrate gut and the tegument of this worm possesses a brush border composed of functional microvilli (Lumsden, 1966) and an external mucopolysaccharide coat which adsorbs high molecular weight charged substances (Lumsden, 1972), as well as inorganic ions (Lumsden, 1973; Lumsden and Berger, 1974).

Taylor and Thomas (1968) found an enhancement of amylase activity in the presence of living tapeworms. Read (1973) supported the findings of these authors when he reported a marked increase in the activity of pancreatic α-amylase in the presence of *H. diminuta*. He concluded that this tapeworm adsorbed pancreatic α-amylase onto its epithelial surface. This was supported by the following findings: 1) maximum relative increase in amylase activity was attained at low enzyme concentrations; 2) the increase in activity is reversed by washing the worms; and 3) high molecular weight polycations partially block the effect. Read suggested that adsorption of amylase onto the epithelial surface of *Hymenolepis* may stabilize the enzyme in a configuration that favors the catalytic activity of amylase.

Ruff, Uglem, and Read (1973) found no interactions between the acanthocephalan, *M. dubius*, and pancreatic trypsin, α-chymotrypsin, β-chymotrypsin, or lipase. This worm did not appear to possess intrinsic proteases or lipases. Intact *M. dubius*, however, freshly removed from the host, had amylolytic activity. It was determined that this amylase activity was of host origin. Borgström et al. (1957) and Goldberg et al. (1968) showed that, in the presence of intact mammalian mucosa, proteolytic enzymes of pancreatic origin are inactivated. Reichenbach-Klinke and Reichenbach-Klinke (1970) reported the inactivation of trypsin in the presence of the tapeworm *Proteocephalus longicollis*. Pappas and Read (1972c) found that intact *H. diminuta* inactivated trypsin when incubated with this protein. These authors postulated that a trypsin inactivator, possibly associated with the glycocalyx of *Hymenolepis*, was highly labile and was detached from the surface of worms after combining with trypsin, thus exposing fresh inactivators. Pappas and Read (1972d) offered the same mechanism to account for the inactivation of α- and β-chymotrypsin by intact *H. diminuta*. Ruff, Uglem, and Read (1973) reported inactivation of pancreatic lipase in the presence of intact *Hymenolepis diminuta*; but further characterization of this inhibitory activity prompted the authors to postulate that the enzyme underwent a loose, transitory attachment to the worm surface by weak bonding, perhaps involving van der Waal forces, resulting in inhibition of activity due to 1) enzyme stabilization in a catalytically unfavorable configuration, or 2) blockage of the "active" sites on the lipase molecule.

Rothman (1966) and Lumsden et al. (1968) conducted cytochemical studies that demonstrated the presence of phosphatases localized in or on the tegumentary brush border of *H. diminuta*. Arme and Read (1970) used biochemical techniques to indicate the presence of phosphatase activity on the surface of *H. diminuta* involved in the hydrolysis of hexose diphosphates. These authors stated, "It has become increasingly clear that the surface of *Hymenolepis*, and probably other tapeworms, should be regarded as a digestive-absorptive structure." Dike and Read (1971a, b) further characterized surface phosphatase activity in *H. diminuta*. They found that the intrinsic tegumentary phosphohydrolase activity occurred at the interface of the worm and the ambient medium. Evidence was provided which indicated that the surface phosphohydrolases of *H. diminuta* are functionally distinct but spatially proximal to the separate system involved in transporting monosaccharides. These data supported the hypothesis that absorption of glucose-6-phosphate occurs in two steps. The first step involves the hydrolysis of glucose-6-phosphate by a surface phosphohydrolase, and the second step is the mediated absorption of the glucose liberated in the first step. Further evidence was provided by the accumulation of glucose in media containing glucose-6-phosphate and an inhibitor of glucose transport. From additional work it was concluded that the hydrolase and the hexose transport systems were located close to one another, between, or at the base of, the microvilli on the surface of the worm.

An intrinsic, membrane-bound ribonuclease (RNase) was demonstrated in *H. diminuta* (Pappas, Uglem, and Read, 1973a). Properties of host pancreatic RNase and *H. diminuta* RNase were compared. Marked differences in pH optimum for enzymatic activity, ion sensitivities, and some kinetic parameters were observed between enzymes from the two sources.

In Read's laboratory, Uglem et al. (1973) observed surface aminopeptidase (APase) activity in the acanthocephalan *M. dubius*. Study of the interactions between the system for the active transport of leucine and the products of leucyl-leucine hydrolysis implied a spatial arrangement between the APase and the leucine transport locus that conferred a kinetic advantage for absorption of the amino acids liberated. Cystocanth larvae of *M. dubius* had no APase activity. Following a thirty-minute exposure of worms to certain surface active agents, however, APase activity was found in larvae.

The scientific endeavors of Clark P. Read, summarized above, provided a large amount of information on the physiology of the vertebrate gut and of its helminth parasites. More importantly, by integrating the prodigious accumulation of data emanating from his and his students' laboratories in these two areas, he significantly advanced our understanding of the molecular basis of changes in the intestinal environment resulting from interactions between hosts and parasites.

Clark Read will be remembered as an outstanding individual and as "the most famous and influential American parasitologist of his age and period" (Simmons, 1974). His short scientific career was marked by the highest level of research productivity.

Present and future generations of investigators in parasitology, intestinal physiology, and symbiosis have suffered an irreparable loss with the passing of Clark P. Read. At the same time, they profit from a legacy that should inspire further advances in these areas.

REFERENCES CITED

Arme, C. and C. P. Read
  1968 Studies on membrane transport. II. The absorption of acetate and butyrate by *Hymenolepis diminuta* (Cestoda). Biological Bulletin **135**:80-91.

  1969 Fluxes of amino acids between the rat and a cestode symbiote. Comparative Biochemistry and Physiology **29**:1135-1147.

  1970 A surface enzyme in *Hymenolepis diminuta* (Cestoda). Journal of Parasitology **56**:514-516.

Bell, L. G.
  1962 Polysaccharide and cell membranes. Journal of Theoretical Biology **3**:132-133.

Borgström, B., A. Dahlquist, G. Lundh, and J. Sjövall
  1957 Studies of intestinal digestion and absorption in the human. Journal of Clinical Investigation **36**:1521-1536.

Chappell, L. H., H. P. Arai, S. C. Dike, and C. P. Read
  1970 Circadian migration of *Hymenolepis* (Cestoda) in the intestine. I. Observations on *H. diminuta* in the rat. Comparative Biochemistry and Physiology **34**:31-46.

Chappell, L. H., C. Arme, and C. P. Read
  1969 Studies on membrane transport. V. Transport of long chain fatty acids in *Hymenolepis diminuta* (Cestoda). Biological Bulletin **136**:313-326.

Crane, R. K.
  1960 Intestinal absorption of sugars. Physiological Reviews **40**:789-825.

  1965 Na^+—dependent transport in the intestine and other animal tissues. Federation Proceedings **24**:1000-1006.

  1967 Structural and functional organization of an epithelial cell brush border. *In* Intracellular Transport. K. B. Warren, ed. New York: Academic Press.

Dike, S. C. and C. P. Read
  1971a Tegumentary phosphohydrolases of *Hymenolepis diminuta*. Journal of Parasitology **57**:81-87.

  1971b Relation of tegumentary phosphohydrolase and sugar transport in *Hymenolepis diminuta*. Journal of Parasitology **57**:1251-1255.

Eichholz, A.
1967 Structural and functional organization of the brush border of intestinal epithelial cells. III. Enzymic activities and chemical composition of various fractions of Tris-disrupted brush borders. Biochimica et Biophysica Acta 135:475-482.

Eichholz, A. and R. K. Crane
1965 Studies on the organization of the brush border in intestinal epithelial cells. I. Tris disruption of isolated hamster brush borders and density gradient separation of fractions. Journal of Cell Biology 26:687-691.

Fieser, L. G. and M. Fieser
1959 Steroids. Third edition. New York: Reinhold Publishing Co.

Fisher, F. M., Jr. and C. P. Read
1971 Transport of sugars in the tapeworm *Calliobothrium verticillatum*. Biological Bulletin 140:46-62.

Ginger, C. D. and D. Fairbairn
1966 Lipid metabolism in helminth parasites. II. The major origins of the lipids of *Hymenolepis diminuta* (Cestoda). Journal of Parasitology 52:1097-1107.

Goldberg, D. M., R. Campbell, and A. D. Roy
1968 Binding of trypsin and chymotrypsin by human intestinal mucosa. Biochimica et Biophysica Acta 167:613-615.

Harris, B. G. and C. P. Read
1968 Studies on membrane transport. III. Further characterization of amino acid systems in *Hymenolepis diminuta* (Cestoda). Comparative Biochemistry and Physiology 26:545-552.

1969 Factors affecting protein synthesis in *Hymenolepis diminuta* (Cestoda). Comparative Biochemistry and Physiology 28:645-654.

Hopkins, C. A. and L. L. Callow
1965 Methionine flux between a tapeworm (*Hymenolepis diminuta*) and its environment. Parasitology 55:653-666.

Ito, S.
1969 Structure and function of the glycocalyx. Federation Proceedings 28:12-25.

Johnson, C. F.
1967 Disaccharidase: localization in hamster intestine brush borders. Science 155:1670-1672.

Kilejian, A.
1966 Permeation of L-proline in the cestode, *Hymenolepis diminuta*. Journal of Parasitology **52**:1108-1115.

Kilejian, A., C. D. Ginger, and D. Fairbairn
1968 Lipid metabolism in helminth parasites. IV. Origins of the intestinal lipids available for absorption by *Hymenolepis diminuta* (Cestoda). Journal of Parasitology **54**:63-68.

Laws, G. F. and C. P. Read
1969 Effect of the amino carboxy group on amino acid transport in *Hymenolepis diminuta* (Cestoda). Comparative Biochemistry and Physiology **30**:129-132.

Lumsden, R. D.
1965 Cytological studies on the absorptive surfaces of cestodes. Ph.D. dissertation, Rice University, Houston, Texas.

1966 Cytological studies on the absorptive surfaces of cestodes. I. The fine structure of the strobilar integument. Zeitschrift für Parasitenkunde **27**:355-382.

1972 Cytological studies on the absorptive surfaces of cestodes. VI. Cytochemical evaluation of electrostatic charge. Journal of Parasitology **58**:229-234.

1973 Cytological studies on the absorptive surfaces of cestodes. VII. Evidence for the function of the tegument glycocalyx in cation binding by *Hymenolepis diminuta*. Journal of Parasitology **59**:1021-1030.

Lumsden, R. D. and B. Berger
1974 Cytological studies on the absorptive surfaces of cestodes. VIII. Phosphohydrolase activity and cation adsorption in the tegument brush border of *Hymenolepis diminuta*. Journal of Parasitology **60**:744-751.

Lumsden, R. D., G. Gonzalez, R. R. Mills, and J. M. Viles
1968 Cytological studies on the absorptive surfaces of cestodes. III. Hydrolysis of phosphate esters. Journal of Parasitology **54**:524-535.

MacInnis, A. J., F. M. Fisher, Jr., and C. P. Read
1965 Membrane transport of purines and pyrimidines in a cestode. Journal of Parasitology **51**:260-267.

Matthews, D. W.
1967 Absorption of water soluble vitamins. British Medical Bulletin **23**:258-262.

McCracken, R., R. D. Lumsden, and Clayton R. Page, III
1975 Sodium-sensitive nucleoside transport by *Hymenolepis diminuta*. Journal of Parasitology **61**:999-1005.

Miller, D. and R. K. Crane
1961 The digestive function of the epithelium of the small intestine. II. Localization of disaccharide hydrolysis in the isolated brush border portion of intestinal epithelial cells. Biochimica et Biophysica Acta **52**:293-298.

Nachlas, M. M., B. Morris, D. Rosenblatt, and A. M. Seligman
1960 Improvement in the histochemical localization of leucine aminopeptidase with a new substrate, L-leucyl-4-methoxy-2-naphthylamide. Journal of Biophysical and Biochemical Cytology **7**:261-264.

Nasset, E. S.
1962 Amino acids in gut content during digestion in the dog. Journal of Nutrition **76**:131-134.

Nasset, E. S. and J. S. Ju
1961 Mixture of endogenous and exogenous protein in the alimentary tract. Journal of Nutrition **74**:461-465.

Nasset, E. S., P. Schwartz, and H. V. Weiss
1955 The digestion of proteins *in vivo*. Journal of Nutrition **56**:83-92.

Pappas, P. W. and C. P. Read
1972a Thiamine uptake by *Hymenolepis diminuta*. Journal of Parasitology **58**:235-239.

1972b The absorption of pyridoxine and riboflavin by *Hymenolepis diminuta*. Journal of Parasitology **58**:417-421.

1972c Trypsin inactivation by intact *Hymenolepis diminuta*. Journal of Parasitology **58**:864-871.

1972d Inactivation of α- and β-chymotrypsin by intact *Hymenolepis diminuta* (Cestoda). Biological Bulletin **143**:605-616.

Pappas, P. W., G. L. Uglem, and C. P. Read
1973a Ribonuclease activity associated with intact *Hymenolepis diminuta*. Journal of Parasitology **59**:824-828.

1973b The influx of purines and pyrimidines across the brush border of *Hymenolepis diminuta*. Parasitology **66**:525-538.

1974 Anion and cation requirements for glucose and methionine accumulation by *Hymenolepis diminuta* (Cestoda). Biological Bulletin **146**:56-66.

Phifer, K.
1960 Permeation and membrane transport in animal parasites: Further observations on the uptake of glucose by *Hymenolepis diminuta*. Journal of Parasitology **46**:137-144.

Platzer, E. G. and L. S. Roberts
1969 Developmental physiology of cestodes. V. Effects of vitamin deficient diets and host coprophagy prevention on development of *Hymenolepis diminuta*. Journal of Parasitology **55**:1143-1152.

1970 Developmental physiology of cestodes. VI. Effect of host riboflavin deficiency on *Hymenolepis diminuta*. Experimental Parasitology **28**:393-398.

Read, Clark P.
1950 The vertebrate small intestine as an environment for parasitic helminths. Monograph in Biology. The Rice Institute Pamphlet 37, No. 2 (July).

1951 The "crowding effect" in tapeworm infections. Journal of Parasitology **37**:174-178.

1952 Contributions to cestode enzymology. I. The cytochrome system and succinic dehydrogenases in *Hymenolepis diminuta*. Experimental Parasitology **1**:353-362.

1953 Contributions to cestode enzymology. II. Some anaerobic dehydrogenases in *Hymenolepis diminuta*. Experimental Parasitology **2**:341-347.

1956 Carbohydrate metabolism of *Hymenolepis diminuta*. Experimental Parasitology **5**:325-344.

1957 The role of carbohydrates in the biology of cestodes. III. Studies on two species from dogfish. Experimental Parasitology **6**:288-293.

1959 The role of carbohydrates in the biology of cestodes. VIII. Some conclusions and hypotheses. Experimental Parasitology **8**:365-381.

1961 Competitions between sugars in their absorption by tapeworms. Journal of Parasitology **47**:1015-1016.

1966 Nutrition of intestinal helminths. *In* Biology of Parasites. E. J. L. Soulsby, ed. New York: Academic Press. Pp. 101-126.

1967 Carbohydrate metabolism in *Hymenolepis* (Cestoda). Journal of Parasitology **53**:1023-1029.

1970 The microcosm of intestinal helminths. *In* Ecology and Physiology of Parasites. A. M. Fallis, ed. Toronto: University of Toronto Press. Pp. 188-197.

1973 Contact digestion in tapeworms. Journal of Parasitology **59**:672-677.

Read, C. P. and A. Z. Kilejian
1969 Circadian migratory behavior of a cestode symbiote in the rat host. Journal of Parasitology **55**:574-578.

Read, C. P. and K. Phifer
1959 The role of carbohydrates in the biology of cestodes. VII. Interactions between individual tapeworms of the same and different species. Experimental Parasitology **8**:46-50.

Read, C. P. and A. H. Rothman
1957a The role of carbohydrates in the biology of cestodes. I. The effect of dietary carbohydrate quality on the size of *Hymenolepis diminuta*. Experimental Parasitology **6**:1-7.

1957b The role of carbohydrates in the biology of cestodes. II. The effect of starvation on glycogenesis and glucose consumption in *Hymenolepis*. Experimental Parasitology **6**:280-287.

1957c The role of carbohydrates in the biology of cestodes. IV. Some effects of host dietary carbohydrate on growth and reproduction of *Hymenolepis*. Experimental Parasitology **6**:294-305.

1958a The carbohydrate requirement of *Moniliformis* (Acanthocephala). Experimental Parasitology **7**:191-197.

1958b The role of carbohydrates in the biology of cestodes. VI. The carbohydrates metabolized *in vitro* by some cyclophyllidean species. Experimental Parasitology **7**:217-223.

Read, C. P., A. H. Rothman, and J. E. Simmons, Jr.
1963 Studies on membrane transport, with special reference to parasite-host integration. Annals of the New York Academy of Science **113**:154-205.

Read, C. P., E. L. Schiller, and K. Phifer
1958 The role of carbohydrates in the biology of cestodes. V. Comparative studies on the effects of host dietary carbohydrate on *Hymenolepis* spp. Experimental Parasitology **7**:198-216.

Read, C. P. and J. E. Simmons
1962 Competitive effects of amino acid mixtures on the uptake of single amino acids by *Calliobothrium*. Journal of Parasitology **48**:494.

Read, C. P., J. E. Simmons, Jr., J. W. Campbell, and A. H. Rothman
   1960 Permeation and membrane transport in parasitism: Studies on a tapeworm-elasmobranch symbiosis. Biological Bulletin **119**:120-133.

Read, C. P., J. E. Simmons, Jr., and A. H. Rothman
   1960 Permeation and membrane transport in animal parasites: Amino acid permeation into tapeworms from elasmobranchs. Journal of Parasitology **46**:33-41.

Read, C. P., G. L. Stewart, and P. W. Pappas
   1974 Glucose and sodium fluxes across the brush border of *Hymenolepis diminuta* (Cestoda). Biological Bulletin **147**:146-152.

Reichenbach-Klinke, H.-H. and K.-E. Reichenbach-Klinke
   1970 Enzymuntersuchungen an Fischer. II. Trypsin- und α-amylase Inhibitoren. Archiv für Fischereiwissenschaft **21**:67-72.

Rothman, A. H.
   1963 Electron microscopic studies of tapeworms: The surface structures of *Hymenolepis diminuta* (Rudolphi, 1819). Blanchard, 1891. Transactions of the American Microscopical Society **82**:22-30.

   1966 Ultrastructural studies of enzyme activity in the cestode cuticle. Experimental Parasitology **19**:332-338.

Ruff, M. D., G. L. Uglem, and C. P. Read
   1973 Interactions of *Moniliformis dubius* with pancreatic enzymes. Journal of Parasitology **59**:839-843.

Simmons, J. E.
   1974 In Memoriam: Clark Phares Read. Journal of Parasitology **60**:385-387.

Simmons, J. E., Jr., C. P. Read, and A. H. Rothman
   1960 Permeation and membrane transport in animal parasites: Permeation of urea into cestodes from elasmobranchs. Journal of Parasitology **46**:43-50.

Taylor, E. and J. Thomas
   1968 Membrane (contact) digestion in the three species of tapeworm *Hymenolepis diminuta*, *Hymenolepis microstoma* and *Moniezia expansa*. Parasitology **58**:535-546.

Ugelov, A. M.
   1965 Membrane (contact) digestion. Physiological Reviews **45**:555-595.

Uglem, G. L., P. W. Pappas, and C. P. Read

    1973 Surface aminopeptidase in *Moniliformis dubius* and its relation to amino acid uptake. Parasitology **67**:185-195.

Woodward, C. K. and C. P. Read

    1969 Studies on membrane transport. VII. Transport of histidine through two distinct systems in the tapeworm *Hymenolepis diminuta*. Comparative Biochemistry and Physiology **30**:1161-1177.

A UNIQUE TEGUMENTARY CELL TYPE AND UNICELLULAR GLANDS ASSOCIATED WITH THE SCOLEX OF *EUBOTHRIUM CRASSUM* (CESTODA: PSEUDOPHYLLIDEA)

by C. Arme and L. T. Threadgold

ABSTRACT

An electron microscope study of the scolex of adult *Eubothrium crassum* (Cestoda: Pseudophyllidea) has revealed the presence of two types of tegumental cells and two types of unicellular glands. The tegument has the cytological organization characteristic of tapeworms with a distal nucleated region (T1 type tegumental cell) connected to a syncytial surface region containing disc-shaped secretory bodies. In addition, a second tegumental cell type (T2) is present and synthesizes a dense ovoid body. It is connected to the surface syncytium by a narrow cytoplasmic tubule, lying within a deep depression of the tegumental base. This tubule is supported by a ring of microtubules that funnel the secretory bodies into the syncytium, the surface of which is consequently evaginated to varying degrees. The two unicellular glands (G1 and G2) have a similar flask shape and internal morphology. Their necks penetrate the muscle layers and the tegument, to which they are attached by a dense ring and a septate desmosome. The G1 cells synthesize a dense granule with the shape of a flattened oval and the G2 cell type synthesizes oval, mucus-like bodies of various densities, which are usually released *en masse* at the tegumental surface.

INTRODUCTION

Gland cells have been described in the scolex of several species of pseudophyllidean cestodes, and the relevant literature has been summarized by Kwa (1972a, b, c) and Öhman-James (1973). This report describes two types of

---

C. Arme is Lecturer in Zoology at The Queen's University, Belfast. L. T. Threadgold is Reader in Zoology at The Queen's University, Belfast.

unicellular glands and a unique tegumentary cell type, found in the scolex of *Eubothrium crassum*.

Rainbow trout (*Salmo gairdneri*) from the Movanagher Fish Farm, Kilrea, Northern Ireland, have been found to be infected with two species of cestode, *Proteocephalus* sp. and *Eubothrium crassum* (Arme and Ingham, 1972; Ingham and Arme, 1973). The scolex of both species was located at the distal end of the pyloric caeca and, in large worms, the strobila extended posteriorly into the small intestine.

Adult *Eubothrium* were dissected from the pyloric caeca of freshly killed fish into a trout saline (Stokes and Fromm, 1964). Scoleces were fixed for 24 hours in 4% glutaraldehyde buffered to pH 7.4 with Millonig buffer, plus 3% sucrose and 0.5 mM calcium chloride. Specimens were then washed for 24 hours in Millonig buffer at pH 7.4, plus 5% sucrose and 0.5 mM calcium chloride, and post-fixed in 1% osmic acid in Millonig buffer for 1 hour. After dehydration through ethanol and propylene oxide, scoleces were embedded in araldite. Sections were cut on an LKB III ultratome, mounted on bare copper grids, and stained for 5 minutes in alcoholic uranyl acetate and then lead citrate. Sections were viewed on an AEI EM 801 and photographs taken at magnifications of 2-40,000×.

Material for scanning electron microscopy was fixed as above. After dehydration, scoleces were transferred to amyl acetate and dried by critical point substitution in a critical point drier (Polaron Ltd.). The dried specimens were coated with gold-palladium and viewed on a Cambridge scanning electron microscope.

1. Scolex tegument

The tegument of the proglottids of *Eubothrium crassum* has the characteristic morphology and organization that are now well established for cestodes. The tegument of the scolex, however, differs from that of the proglottids in a number of ways. In addition to the primary type of tegumentary cell (T1), which synthesizes the small discoidal bodies typical of the proglottid tegument in most cestode species (Beguin, 1966; Lumsden, 1966a, b), there is a second type of tegumentary cell (T2), which is polymorphic and which ramifies between adjacent cells. The relatively large, approximately oval nucleus has a large nucleolus and dense nucleoplasm, which is mainly euchromatic and which contains small patches of heterochromatin (figure 1). The cytoplasm is dense because of an abundance of free ribosomes and contains granular endoplasmic reticulum (GER), usually intimately associated with Golgi complexes, a small number of mitochondria with lucid matrices and few cristae, some groups of β-glycogen granules and secretory bodies (figure 2). These bodies are derived from Golgi complexes and in their mature state are round to oval in section, although some are irregular, and range from slightly flattened ovals to sausage-shaped. The matrix of these secretory bodies is very dense and usually lies close to the bounding membrane, but in many bodies there is a lucid, crescent-shaped gap between content and membrane, giving the secretion a characteristic appearance.

From the tegumentary cells extend long cytoplasmic tubules, which pass through the muscle and connective tissue layers to join onto the base of the syncytial tegument. The proximal regions of these tubules have dense cytoplasm, organelles, and secretory bodies, but distally the cytoplasm is limited to the periphery, leaving a lucid core with secretory bodies. These secretory bodies may be few, or so many that a localized swollen area of the tubule occurs just below the base of the syncytial tegument. The junction between tubule and the base of the syncytium has an unusual organization (figure 3). The cytoplasmic tubules are lined by a peripheral ring of microtubules, which may extend well into the syncytial cytoplasm. Furthermore, the tubules lie in a depression in the base of the syncytium so that the plasma membrane runs up the tubule and is sharply reflected down, parallel with the tubule surface for some distance, before turning at right angles to run parallel with the tegumentary surface (figure 3). At the point of inflection there is a region of increased density, associated with the inner, cytoplasmic aspect of the tubule membrane. The microtubules channel the secretory bodies and confine them to an area of the syncytium that has a wine-glass shape. This area is evaginated to various degrees, ranging from a slightly raised protrusion to a large bulbous structure connected to the tegument by a narrow neck (figures 4 and 5). The fact that these areas are devoid of surface microtrichs suggests that they may

---

FIGURES 1-5 OVERLEAF

FIG. 1. TEGUMENTAL CELL, Type 2, containing T2 secretory bodies (S). N, nucleus; NU, nucleolus; MU, muscle. × 20,000.

FIG. 2. PART OF A T2 TEGUMENTAL CELL showing a Golgi complex (GC), T2 secretory bodies (S) with crescent shaped lucid areas (arrow) and small vacuoles (V); P, parenchymal cells; MU, muscle. × 50,000.

FIG. 3. JUNCTION BETWEEN CYTOPLASMIC TUBULE (T) from T2 tegumental cell and the tegument (TE), showing microtubules (MT) and secretory bodies (S). Note the density at the junction where the basal plasma membrane is reflected back (arrows). × 60,000.

FIG. 4. SURFACE PROTRUSION connected to the tegument by a narrow neck and containing T2 secretory bodies (S). Discoid T1 type secretory bodies are also present (arrows). × 20,000.

FIG. 5. A SURFACE PROTRUSION containing T2 secretory bodies apparently freed from the tegument. × 13,500.

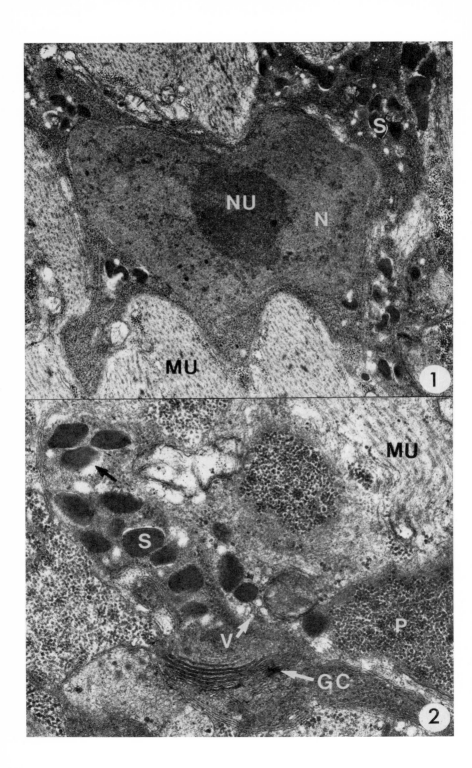

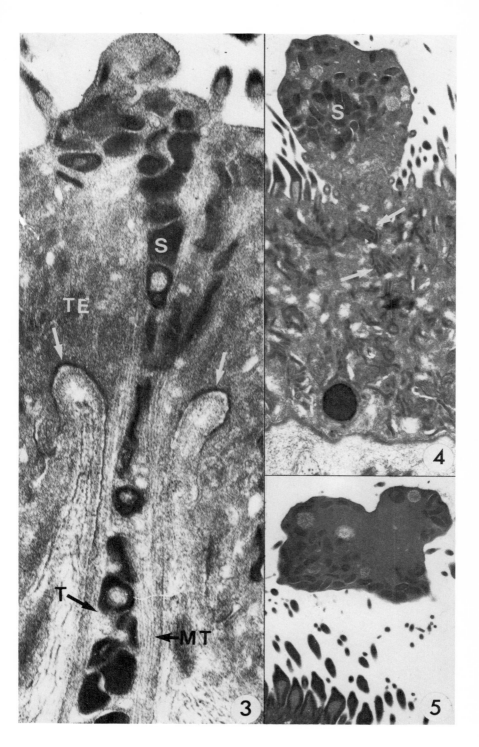

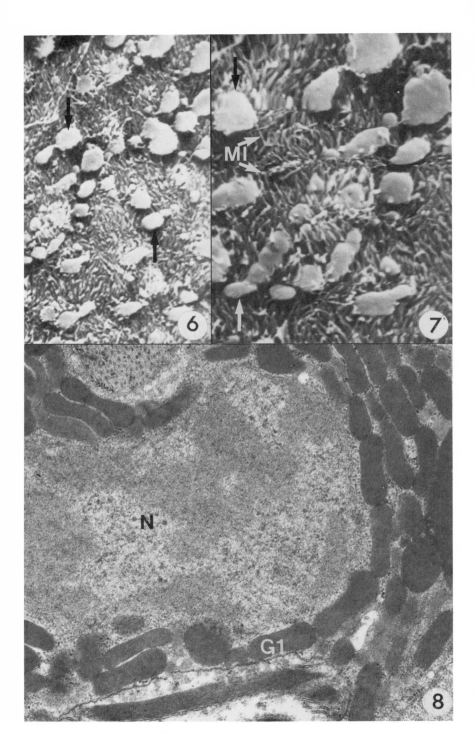

not be permanent features of the scolex tegument. Certain images suggest that the protrusions are eventually pinched off and freed, so that the process resembles apocrine secretion (figure 6). The frequency and heterogeneity of these protruding microtrich-free portions of tegument are revealed by scanning electron microscopy, in which they appear as smooth surfaced, mushroom-like bodies, surrounded by microtrichs (figures 7 and 8). The scanning electron microscope photographs do not show any obviously free bodies on the surface. The tegument adjacent to the protrusions has normal microtrichs, although these are relatively short and well spaced. The cytoplasm adjacent to the protrusion contains secretory bodies identical to those in the protrusions, and such bodies tend to be at right angles to the surface of the tegument. A few mitochondria are also present (figures 3 and 4).

With increasing distance from the scolex there are increasing numbers of disc-shaped bodies, characteristic of T1 tegumentary cells, and decreasing numbers of the large dense bodies (T2 secretion), until the former dominate the syncytial tegument, except in the region of protrusions. Beyond the neck region of the strobila, both protrusions and the large dense bodies are absent, and the microtrichs are longer, more slender, and more closely packed.

2. Unicellular glands

There are two types of unicellular gland in the scolex that overlap in areas of distribution with the T2 tegumentary cell, although they also occur further down the scolex and neck.

The first type of unicellular gland (G1) has a tendency towards being flask-shaped, but this is often modified by indentations from adjacent muscle blocks (figure 9). The nucleus is relatively large, lies basally, and generally follows the outline of the cell, especially laterally. The nucleus contains a large granular nucleolus, wide, ribbon-like masses of heterochromatin, and small, very dense granular masses in a euchromatic nucleoplasm. The cytoplasm is moderately dense with numerous ribosomes and a moderate quantity of GER, which is associated with small Golgi complexes of a few short sacs and many vesicles. Mitochondria are small, round or oval, with a few cristae and lucid matrices (figure 10). The juxta-nuclear cytoplasm is packed with many uniformly dense secretory granules, between which lie numerous β-glycogen granules (figure 9). The secretory granules frequently have a flattened, oval outline and, therefore, resemble a mammalian erythrocyte in shape. Newly synthesized granules tend to have a rounder shape and less dense contents.

FIGS. 6 AND 7. SCANNING ELECTRON MICROGRAPHS of the smooth surfaced protrusions (arrows) and microtrichs (MI). Fig. 6 × 10,400; fig. 7 × 32,000.

FIG. 8. PART OF A GI TYPE UNICELLULAR GLAND CELL with nucleus (N) and secretory bodies (G1). × 32,000.

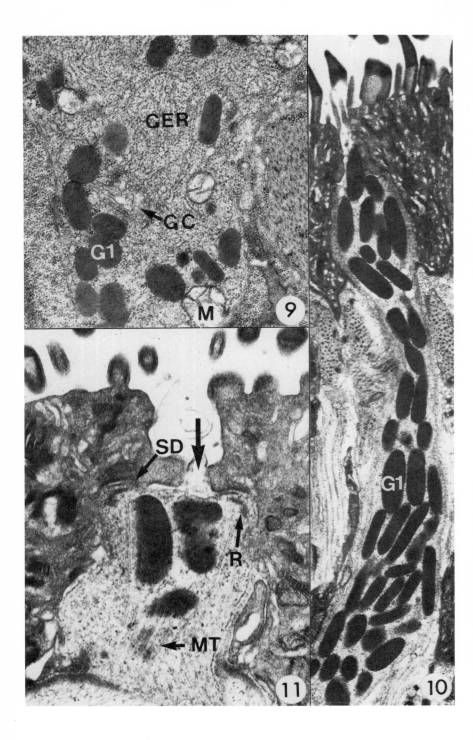

From the distal ends of the cells the cytoplasm extends as a narrow neck or duct which, although originally containing cell organelles and glycogen, soon appears as a large hollow tube with a lucid matrix containing groups of secretory bodies, very small and very dense particles and a few β-glycogen granules. These ducts pass through the muscle layers, interstitial material and basal lamina, to penetrate the tegument itself and open to the exterior between microtrichs (figure 11). In its terminal part, within the tegument, the duct has a peripheral ring of microtubules that terminate in a dense internal ring. The duct is attached to the tegument by a ring-like septate desmosome (figure 12).

The second type of unicellular gland (G2) is, in most respects, morphologically similar to the first type, but its secretory product is distinct. These secretory bodies are generally oval in outline, although they may become quite irregular when the bodies are closely packed. Their content is uniformly granular and ranges from slightly dense to very dense, adjacent bodies often having quite different densities (figure 13). In appearance, therefore, these secretory bodies resemble those from mammalian goblet cells. The distal parts of the ducts of these cells are either almost empty, containing only a few secretory bodies, or are locally swollen by large numbers of closely packed secretory bodies (figure 13). As the ducts approach the tegument they are lined with a peripheral ring of microtubules, and their terminations within the tegument have a dense internal ring and associated ring-like septate desmosome identical to the G1 gland cell described above (figure 14). Within the tegument the duct terminations are often swollen with many secretory bodies, and these masses appear to be released as a single unit.

DISCUSSION

Previous studies on the tegument of adult cestodes have shown that a single type of tegumental cell (T1) is present, which synthesizes a single type of disc-shaped secretory body of varying density, ranging from very electron-dense to light. The evidence suggests that these disc-shaped bodies contain glycoprotein and that their contents are secreted by an eccrine mechanism, when their

---

FIG. 9. PART OF A G1 TYPE GLAND CELL showing the granular endoplasmic reticulum (GER), Golgi complex (GC), secretory bodies (G1) and mitochondria (M). × 20,000.

FIG. 10. TERMINAL PART OF THE DUCT of a G1 type gland cell containing secretory bodies (G1) opening to the exterior. × 15,000.

FIG. 11. HIGHER MAGNIFICATION OF ATTACHMENT of the gland duct to the tegument. Microtubules (MT), septate desmosome (SD), dense ring (R), and the opening to the tegument (arrow) are present. × 40,000.

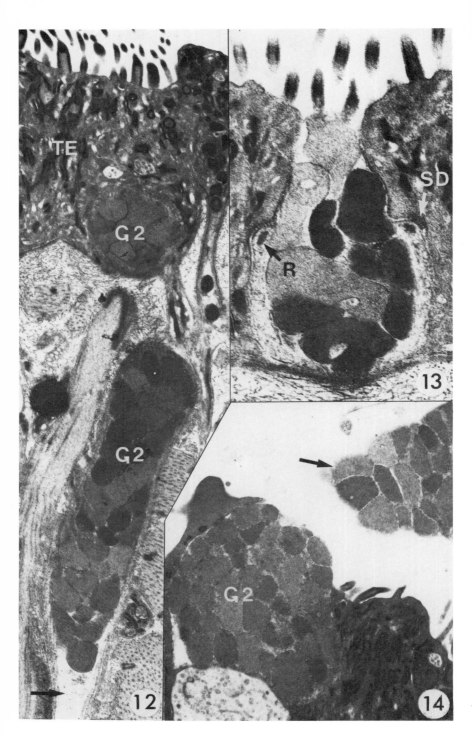

limiting membranes combine with the existing plasma membrane. The presence of vacuoles has also been claimed.

Although adult worms appear to have only one type of secretory body, the tegument of larval pseudophyllideans is reported to contain a number of different types. Kwa (1972c) claimed that two types of granules occurred in the tegument of the sparganum (plerocercoid) larva of *Spirometra erinacei*. The first type of granule was dark and closely packed at the base of the "pit organelle" which had "cilia-like structures" around its opening. The second type of granule was transparent and numbers of them were contained in a membrane-bound "packet." Kwa postulated that the transparent granules were synthesized "in the tegumental cells and then transported as a discrete packet through the cytoplasmic extensions into the distal cytoplasm and eventually released at the surface." Since both types of granule appear to be separated from the cytoplasm by one or two membranes (Kwa, 1972c, figures 2, 6), and open to the exterior, it is extremely unlikely that they are intra-tegumental. The morphology of these granules and their appearance in the zone of the distal tegument is very similar to the swollen terminal parts of the ducts of the two types of unicellular gland described in this paper. In addition, the "pit-organelle" with its packed dark granules resembles to a remarkable degree the pore region of the gland cell type in the scolex of *Diphyllobothrium ditremum* described by Öhman-James (1973). We suggest that Kwa (1972c) has misinterpreted these structures in *Spirometra erinacei*, and that the packets of transparent granules below the muscle layers and in the tegumental cells are, in reality, cross sections of the ducts and cell bodies of unicellular glands; it would seem desirable to examine these structures further. In plerocercoids of *Ligula intestinalis* (Charles and Orr, 1968) there are, in addition to the disc-shaped bodies, striated crystalline bodies, ovoid bodies with granular contents, and vacuolate vesicles; in *Schistocephalus solidus* plerocercoids, there are disc-shaped, crystalline, and vacuolate bodies (Charles and Orr, 1968; Morris and Finnegan, 1969) and in the procercoid larva of *Diphyllobothrium latum* are found disc-shaped and lamellate bodies, the latter resembling myelin figures (Bråten, 1968). The above studies, however, do not furnish evidence concerning whether the different secretory bodies are synthesized by a single type, or a number of different types, of tegumentary cell. Furthermore it is not always clear whether the different types of

FIG. 12. PART OF THE DISTAL PORTION OF THE DUCT of a G2 type gland cell, showing the duct (arrow), secretory bodies (G2) and tegument (TE). × 12,600.

FIG. 13. HIGHER MAGNIFICATION showing the attachment of the duct of a G2 type gland cell to the tegument. The septate desmosome (SD) and dense ring (R) are present. × 30,000.

FIG. 14. THE SWOLLEN OPENING OF A DUCT of a G2 type gland cell is present within the tegument (G2), but a mass of secretory granules appears to be free of the surface (arrow). × 15,000.

secretory bodies described by the above authors in larval cestodes are in reality different, since some of the separate types are of approximately equal size and differ only in the density of their contents. In adult worms the contents of the single disc-shaped body may vary from very dense to almost completely empty, and this variation is especially evident if fixation times are of short duration. The plane of the section may also affect both the shape of a secretory body and the density of its contents. Despite these possibilities of confusion, however, at least two types of secretory bodies appear to have been demonstrated in *Ligula, Schistocephalus,* and *Diphyllobothrium* larvae.

It seems, therefore, that this paper represents the first record of the presence in cestodes of two types of tegumentary cell, T1 and T2, synthesizing clearly separate secretory bodies: disc-shaped (T1 secretion) and ovoid (T2 secretion). Furthermore, the T2 secretion is apparently secreted by an apocrine mechanism, as opposed to an eccrine system for the secretion from T1 cells.

The organization in *Eubothrium* of the junction between the tubule from the T2 tegumental cell and the base of the distal tegument is also unusual, as is the presence of a ring of microtubules in this region, which project well into the distal cytoplasm. The ring of microtubules appears to funnel the migrating secretory granules and distal cytoplasm, so that they form a protrusion at the surface that may possibly be freed as a globular structure containing many secretory bodies. Such protrusions are numerous over the entire scolex, but their function remains unclear.

Possible roles for secretions produced by larval pseudophyllideans have been discussed by Öhman-James (1973) and Kwa (1972c). Previous suggestions for their function have included the production of proteolytic enzymes, possibly to assist in the migration of the larvae or for extracorporeal digestion or for protection of the parasite against the activity of host enzymes. Cytochemical tests on gland cells in *Diphyllobothrium ditremum* (Öhman-James, 1973) were negative for a variety of enzymes tested, and only positive after the Periodic acid-Schiff reaction and several tests for proteins. Kwa (1972b) demonstrated proteolytic activity associated with the tegument of *Spirometra mansonoides.* It is not possible, however, to deduce from Kwa's experiments whether the protease was secreted by the worm or whether there was an intrinsic membrane-bound protease on the surface of the tegument. Dubovskaya (1970) has claimed that her studies on *Bothriocephalus scorpii* indirectly demonstrate the presence of proteases in the tegument of this parasite. Another possible explanation for the observed proteolytic activity in the tegument of the pseudophyllideans described above is that it results from the surface adsorption of host enzymes, as has been described for *Hymenolepis diminuta* (Pappas and Read, 1972a and b). In this case, however, trypsin and α- and β-chymotrypsin were inactivated in the presence of the tapeworm.

The role of the secretory material produced by *Eubothrium crassum* is at present unknown. Possible functions based on the suggestions described above should, however, be amenable to experimental investigation.

REFERENCES CITED

Arme, C. and L. Ingham
  1972 *Proteocephalus sp.* in rainbow trout, *Salmo gairdneri*: a new host record for the Palaearctic. Irish Naturalists Journal 17:241-242.

Beguin, F.
  1966 Étude au microscope electronique de la cuticle et des structures associées chez quelques cestodes. Essai d'histologie comparée. Zeitschrift für Zellforschung und mikroskopische Anatomie 72:30-46.

Bråten, T.
  1968 An electron microscope study of the tegument and associated structures of the procercoid of *Diphyllobothrium latum*. Zeitschrift für Parasitenkunde 30:95-103.

Charles, G. H. and T. S. C. Orr
  1968 Comparative fine structure of outer tegument of *Ligula intestinalis* and *Schistocephalus solidus*. Experimental Parasitology 22:137-149.

Dubovskaya, A. Ya.
  1970 On the possibility of consumption of proteins by the fish cestode *Bothriocephalus scorpii*, Gelan. Voprosy Morskoi Parazit. Kiev. Izdat "Naukova Dumka." Pp. 21-24.

Ingham, L. and C. Arme
  1973 Intestinal helminths in rainbow trout, *Salmo gairdneri* (Richardson): Absence of effect on nutrient absorption and fish growth. Journal of Fish Biology 5:309-313.

Kwa, B. H.
  1972a Studies on the sparganum of *Spirometra erinacei*: I. The histology and cytochemistry of the scolex. International Journal for Parasitology 2:23-28.

  1972b Studies on the sparganum of *Spirometra erinacei*: II. Proteolytic enzyme(s) in the scolex. International Journal for Parasitology 2:29-33.

  1972c Studies on the sparganum of *Spirometra erinacei*: III. The fine structure of the tegument in the scolex. International Journal for Parasitology 2:35-43.

Lumsden, R. D.
  1966a Cytological studies on the absorptive surfaces of cestodes. I. The fine structure of the strobilar integument. Zeitschrift für Parasitenkunde **27**:355-382.

  1966b Cytological studies on the absorptive surfaces of cestodes. II. The synthesis and intracellular transport of protein in the strobilar integument of *Hymenolepis diminuta*. Zeitschrift für Parasitenkunde **28**:1-13.

Morris, G. P. and C. V. Finnegan
  1969 Studies of the differentiating plerocercoid cuticle of *Schistocephalus solidus*. II. The ultrastructural examination of cuticle development. Canadian Journal of Zoology **47**:957-964.

Öhman-James, C.
  1973 Cytology and cytochemistry of the scolex gland cells in *Diphyllobothrium ditremum* (Creplin, 1825). Zeitschrift für Parasitenkunde **42**:77-86.

Pappas, P. W. and C. P. Read
  1972a Trypsin inactivation by intact *Hymenolepis diminuta*. Journal of Parasitology **58**:864-871.

  1972b Inactivation of α- and β-chymotrypsin by intact *Hymenolepis diminuta* (Cestoda). Biological Bulletin **143**:605-616.

Stokes, R. M. and P. O. Fromm
  1964 Glucose absorption and metabolism by the gut of rainbow trout. Comparative Biochemistry and Physiology **13**:53-69.

AMINO ACID POOLS OF *SCHISTOSOMA MANSONI* AND MOUSE HEPATIC PORTAL SERUM

by Harold L. Asch

ABSTRACT

Analyses were made of the free amino acid pools of adult male and female worms and worm pairs of *Schistosoma mansoni* and of the host (mouse) hepatic portal blood. Twenty-five identified and 10-13 unidentified ninhydrin-positive compounds were observed in the parasites. Most of the host blood amino acids were present at concentrations near the respective transport constants (K_t) for the parasite's uptake systems. Comparison of amino acid concentrations within worms to those present in hepatic portal blood suggests that several amino acids, especially glutamate, may be actively concentrated by the parasites. The results are discussed in terms of the physiology and biochemistry of the host-parasite relationship.

INTRODUCTION

Despite the rapidly growing body of information on the biochemistry and physiology of adult schistosomes, their nutritional requirements and mechanisms for obtaining nutrients are poorly understood. Therefore, a series of studies was undertaken to analyze the mechanisms by which these parasites absorb amino acids (Asch and Read, 1975a and b). In the course of these studies, the concentrations of free amino acids in worms and in the host mouse hepatic portal blood were determined.

MATERIALS AND METHODS

A Puerto Rican strain of *Schistosoma mansoni* was maintained in *Biomphalaria glabrata* (PR strain) as described previously (Asch and Read, 1975b). Mice were fed Purina chow *ad libitum*. At seven weeks post-exposure,

---

Harold Asch is a Postdoctoral Fellow in Biochemistry at Baylor College of Medicine.

mice were anesthetized with chlorobutanol (Hunter, 1960) and prepared for perfusion. The perfusion technique was that of Radke et al. (1962) as modified by Pappas and Asch (1972). To assay the amino acids following perfusion, 400-600 worms were rinsed thoroughly and allowed to sit overnight in 70% ethanol. The ethanol was removed, the worms were dried at 80° C for 15 minutes, and their protein content was determined by the method of Lowry et al. (1951). The ethanol was partitioned against three volumes of acidified chloroform and the water layer evaporated at 56° C to 0.3 ml under nitrogen. Amino acids in the sample were measured using a Technicon NC-1 Amino Acid Analyzer, with norleucine as internal standard. The techniques used did not account for potential non-physiological conversions such as glutamine to glutamate or cystine to cysteic acid. The concentrations of amino acids were expressed in terms of ml of tissue water and were calculated from the protein determinations. The ratio of wet weight:dry weight:alcohol extracted dry weight:total protein was 1.00:0.27:0.22:0.20 (P. Pappas, unpublished data; Asch, unpublished data).

For analysis of free amino acids in mouse hepatic portal blood, each mouse was anesthetized with chlorobutanol, the abdomen opened and the portal vein exposed. A small incision was made in the vein and the effusing blood was drawn into a heparinized syringe. Blood (0.4 ml) from one to three mice was centrifuged in a Microfuge (Beckman-Spinco, Palo Alto, California). The plasma was then deproteinized by adding 20 μliters of 50% trichloroacetic acid (TCA). The denatured protein was washed three times with 3% TCA and the washes combined with the supernatant. The preparation was partitioned against acidified chloroform; the water phase was evaporated under nitrogen at 56° C to 0.3 ml and then processed in the amino acid analyzer as described above. No difference was noted in plasma amino acid levels in the mouse when cervical fracture replaced anesthetization or when infected and uninfected mice were compared. Because anesthetization was more convenient and the distended portal veins of infected mice were more accessible, these techniques were employed routinely. Worms and mouse blood were collected between 9:00 A.M. and 11:00 A.M. from mice 7-8 weeks post-infection.

RESULTS

The free amino acid content of mouse hepatic portal plasma (four assays), male worms (two assays) and worm pairs (one assay) are shown in table 1. Cysteic acid, taurine, and urea varied a great deal in their concentration. The variation was due probably to the coincidental elution of these compounds with unidentified ninhydrin-positive compounds and, in the case of cysteic acid, possible contributions from oxidation of cystine and cysteic acid. Although there was some variation in the absolute amounts of any one amino acid from one assay of portal plasma to another, the ratio of a given

TABLE 1
AMINO ACID POOLS OF *S.MANSONI* MALES AND WORM PAIRS AND OF DEPROTEINIZED MOUSE HEPATIC PORTAL PLASMA

COMPOUND	HUMAN PLASMA[1] (Peripheral)	MOUSE PLASMA[2] (Portal)	WORMS[3] Males	Males	Pairs
Aspartate	0.004	0.035 ± .016	ND	0.634	1.108
Threonine	0.127	0.367 ± .126	0.995	0.893	1.400
Serine	0.113	0.198 ± .060	0.748	0.754	1.167
Glutamate	0.051	0.097 ± .013	4.973	3.820	4.192
Proline	0.201	0.178 ± .031	1.283	1.090	1.175
Citrulline	0.021	0.097 ± .020	TRACE	0.126	0.067[r]
Glycine	0.294	0.361 ± .047	0.950	1.060	1.350
Alanine	0.263	0.629 ± .058	3.779	3.990	3.983
Valine	0.244	0.184 ± .073	0.236	0.285	0.380
Cystine	—	0.056 ± .026	TRACE	TRACE	TRACE
Methionine	0.023	0.026 ± .003	0.100[r]	0.100[r]	0.119[r]
Isoleucine	0.069	0.081 ± .015	0.121	0.140	0.207
Leucine	0.128	0.165 ± .040	0.255	0.330	0.454
Tyrosine	0.055	0.063 ± .016	0.140	0.155	0.143
Phenylalanine	0.054	0.069 ± .019	0.141	0.155	0.213
Ornithine	0.051	0.106 ± .035	0.099	0.100	0.228
Lysine	0.171	0.336 ± .082	0.615	0.397	0.706
Tryptophan	—	TRACE	0.020[r]	TRACE	0.038[r]
Histidine	0.060	0.141 ± .052	0.215	0.206	0.366
Arginine	0.089	0.048 ± .009	ND	0.080	0.021[r]
Ammonia	—	0.208 ± .029	0.307	0.080[r]	0.247
γ-aminobutyrate	—	0.015 ± .003	0.059	0.030[r]	0.046[r]
Cysteic acid	—	0.050 ± .026	0.310	0.130	4.460
Taurine	0.065	0.622 ± .197	3.914	0.260	1.867
Urea	—	5.004 ± 1.856	7.261	ND	ND

1. Obtained by averaging data from tables compiled by Diem and Lentner (1970), p. 574.
2. Mean of four determinations ± S.E. Values given as mM.
3. Single determinations. Values given as mM, based on calculation of worm water content from protein assay.
r. Rough estimate (due to small peaks).
ND. Not discernible (due to unresolvable peaks).

amino acid concentration to that of any other remained fairly constant. In addition to the 25 compounds listed, five to seven unidentified ninhydrin-positive compounds were detected in portal plasma. For the parasite, the same 25 compounds were identified in all three assays. Additionally, the numbers of unidentified ninhydrin-positive compounds found in males and worm pairs were ten and thirteen, respectively. Most of the parasite amino acids were present at concentrations similar to those in the host serum.

DISCUSSION

Robinson (1961) found thirteen free amino acids in *S. mansoni* adults but did not report any unidentified ninhydrin-positive compounds. The present report (table 1) indicates a larger number of amino acids plus a substantial number of unidentified ninhydrin-positive materials. Senft (1966) analyzed acid-hydrolyzed, lyophilized worms (containing both free and incorporated amino acids) and reported more than twenty ninhydrin-positive compounds (certain amino acids were omitted because of uncertainty of measurements). If one converts the free amino acid concentrations in *S. mansoni* reported by Chappell (1974) from μmoles/gm dry wt to μmoles/ml worm water, they are quite close to those shown in table 1.

A comparison of the free amino acid levels in worms to those in portal blood suggests that the parasites may concentrate several amino acids. Most notable of these is glutamate, which is approximately forty times higher in worms than in the surrounding blood. A potentially important source of error is the fact that some of the glutamate may have been derived by conversion from glutamine during preparation of the samples for analysis. Other than glutamate, the majority of amino acids appear to be maintained by the worms at concentrations only slightly greater than those found free in the blood. In a single experiment (Asch, unpublished data) no significant change was noted in glutamate concentration within male worms when they were incubated for two minutes in the presence of 0.05 mM glutamate. Thorough accumulation studies will be required to elucidate the extent, if any, of active transport involved in absorption of these amino acids.

The relatively high concentration of glutamate may not be due to specific accumulative transport mechanisms, but instead may result from accumulation of this compound as an end product of metabolism. Schistosomes take up as much as five times their weight in glucose per day (Bueding, 1950) and the high levels of glutamate, alanine, and aspartate may result from transamination of glucose metabolic end products. This hypothesis is supported by the report (Senft, 1963) that 6% of absorbed glucose is converted to alanine. These three amino acids may represent detoxification sinks for amino groups derived from amino acid metabolism, or they may be used as amino group donors for amino acid synthesis. The transaminases of schistosomes have been examined (Garson and Williams, 1957; Huang et al., 1962; Conde del Pino et al., 1968) but their significance in the overall nutrition of these organisms is not known.

The concentrations of free portal plasma amino acids found here (table 1) are similar to those from mouse plasma (unspecified anatomical source, Senft, 1966), and slightly lower than those reported by Page et al. (1972) for cardiac plasma from mice that were hyperinfected, morbid, and starved. They are considerably greater than those reported for human peripheral blood (see Senft, 1966; Diem and Lentner, 1970). The importance of using amino acid concentrations of portal blood when discussing the nutrition of schistosomes must be emphasized. The increase in concentration of free amino acids in the blood following feeding (Van Slyke and Meyer, 1912) is first noticeable in the portal vein, where the highest levels in the blood are attained (Dent and Schilling, 1949; Denton and Elvehjem, 1954). Thus, the schistosomes may be exposed to varying concentrations of amino acids during the course of a day. This is not a source of error in the present studies because food was available *ad libitum* and mice were observed feeding to some degree at all hours, and because both blood and parasites were collected at the same time of day. In man, the natural host of *S. mansoni*, such fluctuations are a probable occurrence, especially in endemically infected populations whose individuals consume poor and restricted diets.

The hemodynamics of the portal blood may also affect the nutritional significance of the concentrations of amino acids available to the schistosomes (see Dent and Schilling, 1949). The total amount of amino acids actually presented to the parasites per unit time may be two or three orders of magnitude greater than the quantity in a few milliliters of stagnant blood. This may affect the rates of transport and metabolism as well as motility of the parasite.

Although the absolute concentrations of amino acids appear to rise following a meal, it is known that the ratio of one amino acid to another is well regulated (Dent and Schilling, 1949; Denton and Elvehjem, 1954). This appears to be consistent with the present findings that, although the absolute portal blood concentrations of a given amino acid varied from one determination to another, its ratio to any other given amino acid remained fairly constant. Most of the blood amino acids are present at concentrations near the respective transport constants (K_t) for the parasite's uptake systems (Asch and Read, 1975b). These findings signal an additional significance of the schistosome amino acid transport systems. The rate of entry of a given amino acid will be controlled by inhibitory interactions among the amino acids that compete with its transport. Since the ratios present in the parasite's milieu are constant, then, although an increase in absolute concentrations may cause higher uptake rates, the ratios of these rates among the amino acids whose uptake is mediated may remain fairly constant. This may be a controlling factor in the worm's biosynthetic capacities. Accordingly, these transport systems may be viewed as parasitic adaptations to homeostatic mechanisms of the host (Read et al., 1963).

ACKNOWLEDGMENTS

This work was carried out in the laboratory of the late Clark P. Read and represents part of the fulfillment of the Ph.D. degree requirements at Rice University. I am especially grateful to Dr. Read for guidance, Dr. P. Pappas for suggestions, and Mr. W. Kitzman for technical assistance. Appreciation is also accorded Drs. S. Bishop and M. Dresden for reviewing the manuscript. Support was received from Grant #3T01A100106-1451 from NIH and Contract #DADA 17-73-3068 from the U. S. Army Research and Development Command.

REFERENCES CITED

Asch, H. L. and C. P. Read
1975a Transtegumental absorption of amino acids by male *Schistosoma mansoni*. Journal of Parasitology **61**:378-379.

1975b Membrane transport in *Schistosoma mansoni*: Transport of amino acids by adult males. Experimental Parasitology **38**:123-135.

Bueding, E.
1950 Carbohydrate metabolism of *Schistosoma mansoni*. Journal of General Physiology **33**:475-495.

Chappell, L.
1974 Methionine uptake by larval and adult *Schistosoma mansoni*. International Journal for Parasitology **4**:361-369.

Conde del Pino, E., A. M. Annexy-Martínez, M. Pérez-Vilar, and A. A. Cintrón-Rivera
1968 Studies in *Schistosoma mansoni*. II. Isoenzyme patterns for alkaline phosphatase, isocitric dehydrogenase, glutamic oxalacetic transaminase, and glucose-6-phosphate dehydrogenase of adult worms and cercariae. Experimental Parasitology **22**:288-293.

Dent, C. and J. Schilling
1949 Studies on the absorption of proteins: the amino acid pattern in the portal blood. Biochemistry **44**:318-355.

Denton, A. and C. Elvehjem
1954 Availability of some amino acids *in vivo*. Journal of Biological Chemistry **206**:449-460.

Diem, K. and C. Lentner, eds.
1970 Scientific Tables. Basle: CIBA-Geigy Ltd.

Garson, S. and J. S. Williams
1957 Transamination in *Schistosoma mansoni*. Journal of Parasitology
43(suppl.):27-28.

Huang, T. Y., Y. H. Tao, and C. H. Chu
1962 Studies on transaminases of *Schistosoma japonicum*. Chinese
Medical Journal 81:79-85.

Hunter, G. W.
1960 The use of anticoagulants and chlorobutanol for the recovery of
adult schistosomes from mice. Journal of Parasitology 46:206.

Lowry, O., N. Rosebrough, A. Farr, and R. Randall
1951 Protein measurement with the Folin phenol reagent. Journal of
Biological Chemistry 193:265-275.

Page, C., F. Etges, and J. Ogle
1972 Experimental prepatent Schistosomiasis mansoni: Quantitative
analyses of proteins, enzyme activity and free amino acids in mouse
serum. Experimental Parasitology 31:341-349.

Pappas, P. and H. Asch
1972 Modification of the Perf-O-Suction technique for schistosome
recovery. International Journal for Parasitology 2:283.

Radke, M., S. Garson, and L. Berrios-Duran
1962 Filtration devices for separating parasites from fluids. Journal of
Parasitology 48:500-501.

Read, C. P., A. Rothman, and J. Simmons
1963 Studies on membrane transport, with special reference to parasite-
host integration. Annals of the New York Academy of Sciences
113:154-205.

Robinson, D.
1961 Amino acids of *Schistosoma mansoni*. Annals of Tropical
Medicine and Parasitology 55:403-406.

Senft, A.
1963 Observations on amino acid metabolism of *Schistosoma mansoni*
in a chemically defined medium. Annals of the New York Academy
of Sciences 113:272-288.

1966 Studies in arginine metabolism by schistosomes. I. Arginine uptake and lysis by *Schistosoma mansoni*. Comparative Biochemistry and Physiology **18**:209-216.

Van Slyke, D. and G. Meyer
    1912 The amino acid nitrogen of the blood. Preliminary experiments on protein assimilation. Journal of Biological Chemistry **12**:399-410.

ACANTHOCEPHALAN DEVELOPMENT: MORPHOGENESIS OF LARVAL *MONILIFORMIS DUBIUS*

by J. E. Byram and Kay W. Byram

ABSTRACT

The nine larval stages of *Moniliformis dubius* as seen in the cockroach intermediate host, *Periplaneta americana,* are illustrated, and detailed tabulations of the events of larval morphogenesis as influenced by temperature, season, photoperiod, and elevated temperature are presented. Significantly increased rates of development were seen in the summer as opposed to winter or fall and under a 12 hours light/12 hours darkness regime versus continuous dark. Morphological anomalies were present in 80% of the cockroaches harboring normal larvae and the number of anomalies was found to decrease significantly as the number of larvae per host increased. High temperature (35° C) invoked a block in morphogenesis, with few normal larvae developing beyond the II* acanthor stage, and resulted in appreciably stimulated host reactions. Attempts to circumvent this effect met with only limited success.

INTRODUCTION

The development of helminth symbionts in their arthropod intermediate hosts is influenced by a variety of environmental parameters, including those acting on the host and subsequently on the symbiont, as well as those which directly affect the symbiont. The larval development of the acanthocephalan *Moniliformis dubius* may be conveniently ordered into a series of morphologically distinct stages (Van Cleave, 1947). A comparison of the data of the early studies of Moore (1946) and of King and Robinson (1967),

---

J. E. Byram is Research Associate in Pathology at Harvard Medical School and Peter Bent Brigham Hospital. Kay W. Byram is pursuing studies in early childhood education at Wheelock College, Boston.

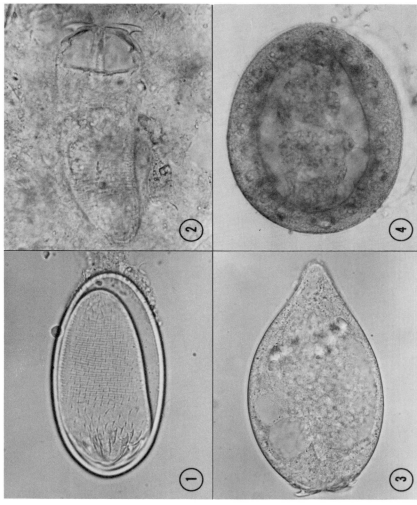

FIGS. 1-14. PHOTOMICROGRAPHS illustrating the various larval stages of *Moniliformis dubius* recovered from the cockroach intermediate host, *Periplaneta americana*, during the course of this study.

FIG. 1. STAGE I* ACANTHOR— isolated from the gut washings of a cockroach incubated 9 days after infection at room temperature (22-24°C). At this point, escape from the egg shell and membranes is imminent. × 565.

FIG. 2. STAGE II* ACANTHOR— seen *in situ* in the gut tissues of a cockroach incubated 11 days after infection at room temperature (22-24°C). × 560.

FIG. 3. STAGE II* ACANTHOR— as recovered from the hemocoel of a cockroach incubated 9 days after infection at 30°C. The rostellar hooks are still present on the anterior end of the larva and few details are apparent in the central nuclear mass. × 385.

FIG. 4. STAGE I ACANTHELLA (EARLY). Initial differentiation of the central nuclear mass into

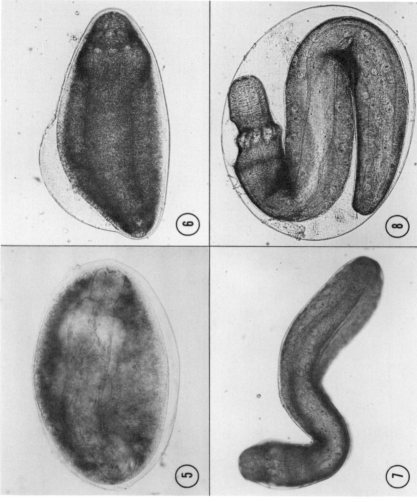

FIG. 5. STAGE I ACANTHELLA (LATE). Organ systems show considerable development but the lemniscal nuclear ring characteristic of the Stage II acanthella is not yet evident. × 95.

FIG. 6. STAGE II ACANTHELLA. The formation of the lemniscal nuclear ring and initial body elongation mark this stage. × 65.

FIG. 7. STAGE III ACANTHELLA. The larval body assumes an elongated cylindrical shape and begins to fold within the enclosing envelope (not seen here). × 55.

FIG. 8. STAGE III ACANTHELLA (LATE) — the characteristic z-shape of this stage. Hook nubs may be seen in the developing proboscis. × 60.

organ systems and formation of the body wall have begun. The rostellar hooks of the acanthor stage are occasionally found on the early acanthella but are typically displaced about one-eighth of the larval circumference from their previous anterior position. × 200.

FIG. 9. STAGE IV ACANTHELLA.
The appearance of the lemnisci,
an elongated body shape and
well developed hooks in the
cytoplasm of the proboscis
tegument typify this stage. × 35.

FIG. 10. STAGE V ACANTHELLA.
Shortening, broadening and
flattening of the larval body is
seen. The proboscis musculature
is complete and the tips of the
hooks protrude from the proboscis surface. × 30.

FIG. 11. STAGE VI ACANTHELLA.
This stage begins with the inver-
sion of the proboscis and neck
and ends with the appearance of
the cystacanth. × 75.

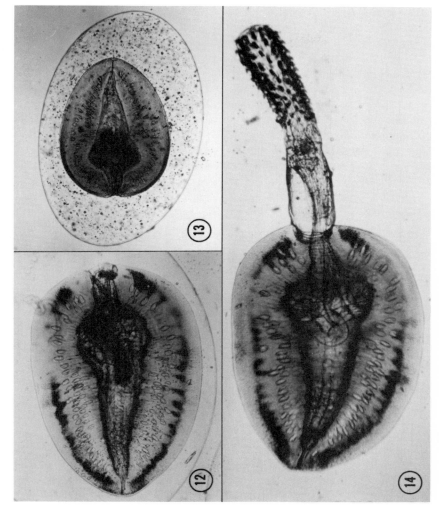

FIG. 12. CYSTACANTH—the infective stage. The body is distinctly compacted, much shorter than the Stage VI acanthella, and compressed, with one surface being flattened or slightly concave. The proboscis and neck remain tightly withdrawn and the cortical nuclei have lost the rounded shape seen in the previous stage. × 45.

FIG. 13. CYSTACANTH. In an occasional cockroach, refractile granules of an undetermined nature were present between the cystacanth surface and the enclosing envelope. × 40.

FIG. 14. CYSTACANTH—proboscis and neck everted as is typical of a newly activated juvenile found in the intestine of the rat definitive host. Cystacanths of this appearance in the cockroach hemocoel are considered to be uninfective. × 55.

and the more recent findings of Lackie (1972a) reveal that ambient temperature governs the rate of larval morphogenesis of this symbiont. King and Robinson (1967) presented data suggesting a similar rate influence by season. Subsequent studies by Robinson and Jones (1971) and Lackie (1972b) revealed that elevated temperatures (above 30° C) markedly altered the normal course of larval morphogenesis. We have found that only by a detailed tabulation of the morphogenetic events in acanthocephalan larval development, such as is seen in King and Robinson (1967, table I) and in Lackie (1972a, table 10), can the trends and subtleties of these events be fully appreciated and analyzed. In this paper, selected aspects of the effects of temperature, season, photoperiod, and temperature stress on the larval morphogenesis of *M. dubius* are examined using such tabulations.

MATERIALS AND METHODS

The acanthocephalan, *Moniliformis dubius* Meyer, 1933, was maintained in the laboratory as described by Byram and Fisher (1973). Adult *Periplaneta americana* were fed eggs taken from gravid female worms 49 to 154 days after infection of the rat definitive host.

Infected roaches were cultured under a variety of temperature and light regimes: 1) room temperature (23° C and 26.5° C) with seasonal illumination, 2) 30° C with 24-hour darkness, 3) 30° C with 12-hour darkness/12-hour light, 4) 35° C with 12-hour darkness/12-hour light, and 5) 30° C with 24-hour darkness to a certain developmental stage (II* acanthor and I acanthella) and transferred to 35° C with 12-hour darkness/12-hour light to completion of the experiment. A saturated atmosphere was insured at 30° C and 35° C.

M. dubius larvae were recovered from *P. americana* by cutting off the head of a roach and flushing out the hemocoel contents with a Pasteur pipette filled with tap water, or with KRTM (Read et al., 1963) when the presence of later developmental stages was expected. The body of the roach was opened and the tissues teased out and searched for any adhering larvae. Counts were made of each larval stage present in an individual roach, following the terminology of King and Robinson (1967). All larval stages and anomaly types were photographed in fresh mounts soon after removal from the cockroach hemocoel, under a Zeiss microscope.

OBSERVATIONS

Even though various photomicrographs and drawings of the larval stages of *Moniliformis dubius* have been published (Moore, 1946; King and Robinson, 1967), a complete collection of illustrations has been lacking. Figures 1–14 show the major features necessary to distinguish each of the nine stages seen in the cockroach intermediate host, *Periplaneta americana,* and, when utilized together with the concise definitions of King and Robinson

(1967), provide a convenient guide for future investigators. The increasing sizes of the various larval stages through the course of development are indicated by the decreasing magnifications of the successive figures. The Stage VI acanthella (figure 11), which previously has not been illustrated, was first recognized by King and Robinson (1967) and clearly can be differentiated from the cystacanth (figures 12–14). These photographs (figures 1–14) emphasize that the cystacanth (so fortuitously named by King and Robinson, as shown by the later studies of Rotheram and Crompton [1972], who found the envelope enclosing the larvae to be indeed a cyst) is the end result of gradual progressive larval development rather than the product of a dramatic transformation or metamorphosis (Van Cleave, 1947).

Rates of Morphogenesis

The development of larval *Moniliformis dubius* in *Periplaneta americana* required less time as the temperature increased (tables 1–6). With seasonal illumination at 23°C (table 1) and 26.5°C (table 2), development in 50-day-old infections had reached 94.6% Stage V and 54.3% Stage VI acanthella respectively. As the result of an accident, development to the cystacanth stage was not completed at 23°C, but was accomplished at 26.5°C in 55-60 days. With 12 hours darkness and 12 hours light, morphogenesis at 30°C was terminated in 35 days (table 6). The developmental rates at these three temperatures are compared graphically in figure 15. These graphic depictions of the morphogenetic rates allow an analysis of the course of development. An estimate of the duration of each stage from the 30°C samples gives the following figures: II* acanthor, 5 days; I acanthella, 3.5 days; II acanthella, 2 days; III acanthella, 2.8 days; IV acanthella, 3.7 days; V acanthella, 4.5 days; and VI acanthella, 5.5 days. Thus, it can be seen that great variations exist in the timing of the development of each stage to the next and that a regression line is of limited use in estimating when a designated stage might appear. The effects of lowering the temperature can be seen in delayed hatching of the acanthor in the cockroach gut and penetration of the gut, and

---

TABLES OVERLEAF

EXPLANATION OF TABLES 1-6. The morphogenesis of larval *Moniliformis dubius* in the cockroach intermediate host, *Periplaneta americana,* under various temperature and light regimes. Each column represents the larvae recovered from one host at the indicated time after infection. The numbers in parentheses (N) at the top of each column show the number of larvae recovered from each cockroach. Anomalies are represented as the percentage of anomalous larvae in each sample. The relative composition by percentage of the morphologically normal larvae in each sample (exclusive of those showing developmental anomalies) is listed under the appropriate larval type.

TABLE 1

Temperature, 23° C; date of infection, 1/28/70; light regime, seasonal. This experiment was terminated at 50 days with 94.7% of the larvae at Stage V acanthella.

(N)	(17)	(64)	(58)	(139)	(18)	(21)	(359)
ANOMALIES	0	0	20.7	7.9	5.6	4.8	0.3
II* acanthor	100	29.7					
I acanthella		70.3	13.0	19.5			
II acanthella			87.0	72.7	47.1	5.0	2.8
III acanthella				7.8	17.6	5.0	0.6
IV acanthella					35.3	40.0	1.4
V acanthella						50.0	94.7
VI acanthella							0.6
	20	25	30	35	40	45	50
				DAYS AFTER INFECTION			

TABLE 2

Temperature, 26.5° C; date of infection, 6/22/70; light regime, seasonal. *a*, anomalies not recorded in day 55 sample.

(N)	(5)	(38)	(25)	(6)	(23)	(22)	(11)	(136)	(275)	(66)
ANOMALIES	0	0	8.0	33.0	4.3	4.5	0	5.2	a	1.5
II* acanthor	100	15.8								
I acanthella		84.2	43.5	25.0						
II acanthella			56.5	75.0	31.8	4.8	9.1	1.6		
III acanthella					36.4			1.6		
IV acanthella					31.8	52.4	9.1	7.8		
V acanthella						42.8	81.8	35.7	2.5	1.5
VI acanthella								53.5	19.3	9.2
cystacanth									78.2	89.3
	15	20	25	30	35	40	45	50	55	60
					DAYS AFTER INFECTION					

TABLE 3

Temperature, 30° C; date of infection, 2/4/70; light regime, 24 hours darkness.

(N)	(26)	(94)	(207)	(337)	(438)	(239)
ANOMALIES	11.5	6.3	1.0	0.3	2.1	1.3
II* acanthor	39.1					
I acanthella	60.9	10.2				
II acanthella		50.0	12.7	1.5	5.4	
III acanthella		5.7	3.9	0.6	2.3	
IV acanthella		34.1	10.7	7.4	11.7	
V acanthella			72.7	15.8	26.3	6.4
VI acanthella				74.7	21.0	9.3
cystacanth					33.3	84.3
	15	20	25	30	35	40
			DAYS AFTER INFECTION			

TABLE 4

Temperature, 30° C; date of infection, 6/30/70; light regime, 24 hours darkness.

(N)	(177)	(134)	(30)	(124)	(9)	(130)
ANOMALIES	27.7	1.5	0	3.2	0	1.5
II* acanthor	3.9					
I acanthella	76.6	0.8				
II acanthella	19.5	0.8				1.6
III acanthella						
IV acanthella		98.4	6.7	1.7		0.8
V acanthella			83.3	25.0	11.1	3.9
VI acanthella			10.0	1.7		5.4
cystacanth				71.6	88.9	88.3
	15	20	25	30	35	39
			DAYS AFTER INFECTION			

TABLE 5

Temperature, 30° C; date of infection, 10/5/70; light regime, 24 hours darkness.

(N)	(101)	(13)	(10)	(89)	(49)	(149)
ANOMALIES	10.9	15.4	10.0	2.2	12.2	5.4
II* acanthor	5.5					
I acanthella	46.7		11.1		2.3	
II acanthella	47.8	36.4	22.2	5.7	7.0	
III acanthella		9.1				
IV acanthella		54.5	55.6	2.3		0.7
V acanthella			11.1	20.7	18.6	5.7
VI acanthella				70.1	32.6	
cystacanth				1.2	39.5	93.6
	15	20	25	30	35	40
			DAYS AFTER INFECTION			

TABLE 6

Temperature, 30° C; date of infection, 10/5/70; light regime, 12 hours light/ 12 hours darkness. *a*, total number of larvae not recorded in day 8 sample.

(N)	a	(40)	(124)	(88)	(16)	(78)
ANOMALIES	0	40.0	2.4	0	6.3	5.1
II* acanthor	100	12.5				
I acanthella		37.5				
II acanthella		50.0	12.4			
III acanthella			7.4			
IV acanthella			79.3	4.5		
V acanthella			0.8	94.3	13.3	
VI acanthella				1.2	66.7	
cystacanth					20.0	100
	8	15	20	25	30	35
			DAYS AFTER INFECTION			

TABLE 7

Probability values (p) for the significance of the difference of the mean developmental stages of the larvae examined at the indicated times when compared on the basis of season (figure 16) and light regime (figure 17). Significance probabilities were computed from the raw data used to construct tables 3-6 using the Wilcoxon two-sample test for two samples, ranked observations, not paired. *, significant.

DAY	15	20	25	30	35	40
2/4 vs. 6/30	p<0.001*	p<0.001*	p<0.02	p<0.001*	p<0.01*	- - - -
6/30 vs. 10/5	p<0.01*	p<0.1	p<0.001*	p<0.001*	p<0.05*	- - - -
2/4 vs. 10/5	p<0.001*	p<0.2	p<0.01*	p<0.9	p<0.1	p<0.2
12/12 L/D vs. 24 D	p<0.9	p<0.2	p<0.001*	p<0.05*	p<0.001*	- - - -

in the duration of each of the subsequent larval stages. In general, when the rates of morphogenesis at 30° C and 26.5° C are compared, the duration of each stage is seen to increase 1.1 to 1.5 times; however, the time spans of the Stage I acanthella and Stage IV acanthella are increased 3.1 and 2.3 times respectively. A similar selective effect of lowering the temperature on these two stages is seen at 23° C.

Tables 3–5 show an apparent seasonal effect on the infections carried out in winter, summer, and fall under constant temperature (30° C) and photoperiod (24 hours darkness) regimes. The rate of development of cystacanths in 35 days varied from a low of 33.3% in winter and 39.5% in fall to a high of 88.9% in summer. The morphogenetic rates at these seasons are compared graphically in figure 16. The mean developmental stages at the indicated points of time in development during each season were then compared statistically (table 7). Since morphogenetic progress was grouped into classes (stages), a nonparametric test of significance, the Wilcoxon two-sample test for two samples, ranked observations, not paired (Sokal and Rohlf, 1969), was utilized. The trends in table 7 are clear. The developmental rates during winter and fall do not differ significantly from one another, but development during the summer proceeds at a significantly higher rate than that during either the winter or the fall. These results support the observation of King and Robinson (1967) that at 27° C development to the cystacanth stage is more rapid in summer (5–7 weeks) than in winter (7–8 weeks).

An apparent effect of photoperiod on two parallel experiments carried out in the fall at 30° C is seen in tables 5 and 6. At 35 days, 100% of the larvae from a 12 hours light/12 hours darkness regime were at the cystacanth stage, while only 39.5% of the larvae under a 24 hour darkness regime were cystacanths. The morphogenetic rates under these two light regimes are depicted in figure

EXPLANATION OF FIGURES 15-17. The rate of morphogenesis of larval *Moniliformis dubius* in the cockroach intermediate host, *Periplaneta americana*, examined on the basis of temperature (figure 15), season (figure 16), and photoperiod (figure 17). Each point on the curves represents the mean developmental stage of the larvae recovered at the indicated time. For the purposes of comparison, each larval stage was assigned an arbitrary value (II* acanthor, 2; I acanthella, 3; II acanthella, 4; etc.) and the developmental means were computed from the raw data used to construct tables 1-6.

FIG. 15. A GRAPHIC COMPARISON of the developmental rates at three different temperatures—30° C (table 6), 26.5° C (table 2) and 23° C (table 1).

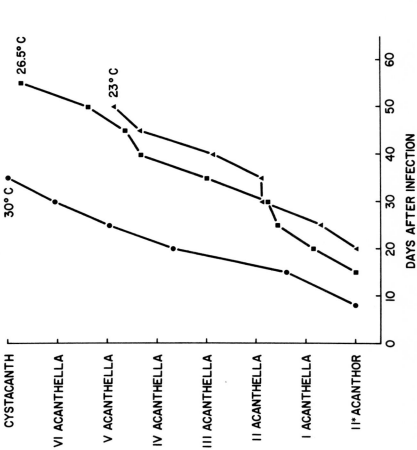

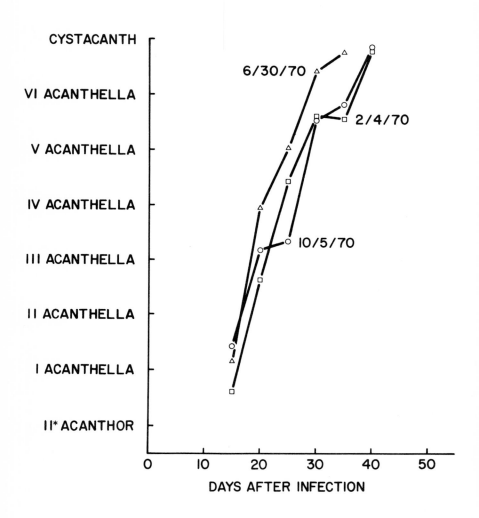

FIG. 16. A GRAPHIC COMPARISON of the developmental rates at three different seasons—winter (table 3), summer (table 4), and fall (table 5). The cockroach hosts were maintained under identical temperature (30° C) and light (24 hours darkness) conditions.

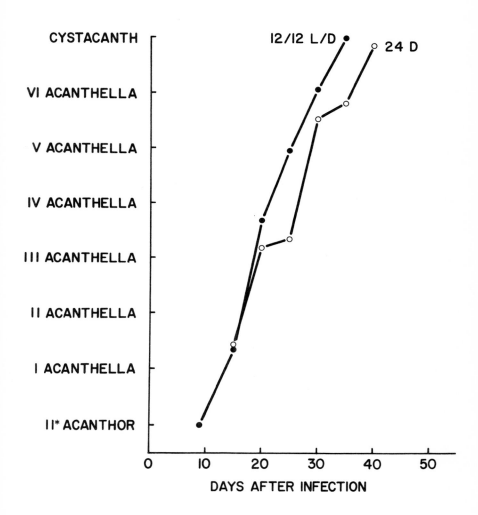

FIG. 17. A GRAPHIC COMPARISON of the developmental rates under two different photoperiod regimes—12 hours light/12 hours darkness (table 6) and 24 hours darkness (table 5). These incubations ran parallel to one another beginning from the date of infection, 10/5/70.

17. As seen in table 7, a comparison of the developmental rates at each point shows that the rates are progressively diverging and that development proceeds at a rate significantly higher under a regime of 12 hours light/12 hours darkness than under one of 24 hours darkness.

Anomalous Development

Of the 3,985 larvae studied under temperature conditions giving rise to normal development (tables 1–4), 174 or 4.37% were observed to be morphological anomalies. In general, the occurrence of anomalies paralleled normal development, with the early stage anomalies being found in the early period of development and later stage anomalies appearing later in development. This presents a puzzle, since no early stage anomalies were seen in the late phase of development. These early stage anomalies must develop to later normal or anomalous stages or be destroyed and cleaned out by a host response. Most anomalies occurred early in development, with the Stage II acanthella accounting for 42% of the total and the sum of Stage I and II acanthellas for 71%. Most later stage anomalies were distributed among the Stage IV (5.75%) and V (11.5%) acanthellas and cystacanths (6.9%). Anomalous development was seen in all stages and included features such as gross distortion of normal shape, blebs, undersized or shrunken and withered larvae, and misshapen proboscides.

The varying numbers of larvae recovered from each cockroach and the differing frequencies of anomalous development suggested the possibility of a crowding effect resulting in these anomalies, as is seen in the development of larval cestodes in the insect intermediate host (Schiller, 1959). This relationship is analyzed in the scatter diagram (figure 18). Since most of the percentages of anomalies fell below 30%, this distribution was normalized by the arcsin transformation (Sokal and Rohlf, 1969). The skew to the right in the frequency distribution of the number of larvae per cockroach was made more symmetrical by plotting these values on a logarithmic scale. The values used in figure 18 were taken from the experiments depicted in tables 1–6 and from one additional experiment run at an average temperature of 28.25° C. When all of the hosts are considered, including those from which no anomalies were recovered (0% anomalies), the regression of y on x approaches 0 ($y = 11.69 - 0.0661x$) and no correlation is seen between percentage of anomalies and larvae/host (correlation coefficient, −0.00375). If one asks whether a relationship exists between the frequency of anomalies and the number of larvae per cockroach *when anomalies are found to occur*, however, the regression of y on x is found to be negative ($y = 28.32 - 7.2703x$) and a strong correlation is evident (correlation coefficient, −0.432). Thus, a significant correlation ($p < 0.01$) exists between the percentage of anomalies and the number of larvae in a cockroach, such that the percentage of anomalies decreases as the number of larvae increases. This effect might be termed a negative crowding effect.

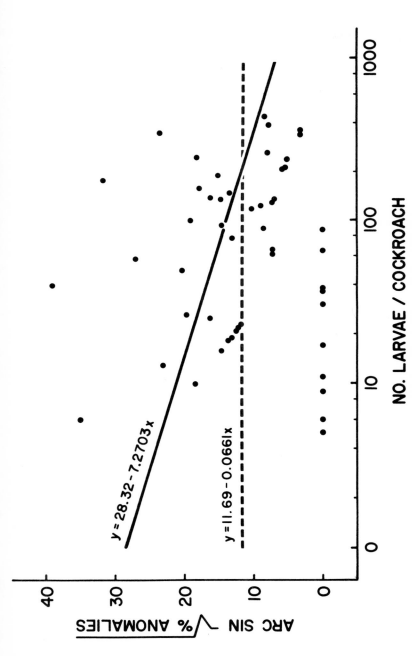

FIG. 18. SCATTER DIAGRAM of the relationship between the percentage of anomalous larvae in a given host and the total number of larvae, normal and anomalous, in that particular host. Regression lines fitted by the method of least squares. The dashed line ($y = 11.69 - 0.0661 x$, N=50) represents all cases including those hosts harboring only larvae showing normal morphology, whereas the solid line ($y = 28.32 - 7.2703 x$, N=40) is restricted to those cases in which anomalous development occurred. See text for further details.

Effects of High Temperature (35° C)

As has been shown by Robinson and Jones (1971) and Lackie (1972b), temperatures of 32-33°C and above markedly alter the normal course of larval development of *M. dubius*. The following experiments were intended to shed some light on this phenomenon.

Morphogenesis at 35°C (figure 20; see figure 19 for explanation) was followed to determine the nature and timing of abnormal development. Stage II* acanthors in the gut tissues appeared normal at day 5. A few Stage I acanthellas were found by day 13; however, development progressed no further though most of the larvae recovered through day 23 were morphologically normal. Thus the first obvious manifestation of anomalous development at high temperature is a suspension of larval morphogenesis. By day 25 most larvae were morphologically abnormal and by day 30 no normal larvae were recovered. Anomalous morphological features at high temperature included those previously listed and, in addition, a series of host reaction effects—host cellular responses, cloudiness, degeneration of the larvae, and melanization—not evidenced at 30°C and below within the time frames of these experiments.

Two further experiments were conducted to attempt to alleviate the morphogenetic block imposed by high temperature. First, the data of Robinson and Jones (1971) suggested that Stage II* acanthors might have difficulty in escaping from the gut tissues into the hemocoel at high temperatures. A possible trapping of emerging acanthors by a host reaction at the gut surface may be inferred from the study of Lackie (1972b). To circumvent this problem, cockroaches were cultured at 30°C until all Stage II* acanthors had migrated into the hemocoel (day 9) and from that point were incubated at 35°C (figure 21).

The last experiment was based on the failure of the morphogenesis of most Stage II* acanthors into Stage I acanthellas at 35°C. If, in the larval development of *M. dubius*, a true metamorphosis occurs, it is in the transformation of the acanthor into the acanthella. To determine whether the major effects of high temperature related to this transformation, cockroaches were cultivated at 30°C until a majority of the larvae had passed this point and were Stage I acanthellas (day 15) and from that point were incubated at 35°C (figure 22).

In neither of these experiments was the morphogenetic block imposed by high temperature successfully evaded. In the first experiment, a slight alteration of the 35°C results was evidenced, and in the second experiment the course of development proceeded somewhat normally through day 25; the frequency of anomalies increased steadily throughout the sampling periods, however, until all larvae at day 35 were morphologically anomalous. In addition to those anomalous features seen at 35°C, a new defect characterized by a series of pseudocoel alterations appeared in these two experiments. The

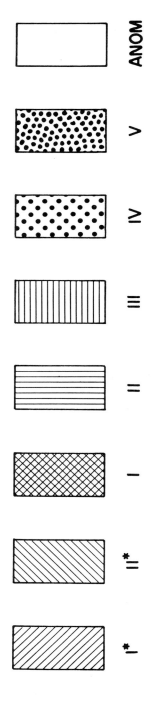

FIG. 19. A KEY TO THE LARVAL STAGES of *Moniliformis dubius* represented in Figs. 20-22. I*, Stage I* acanthor; II*, Stage II* acanthor; I, Stage I acanthella; II, Stage II acanthella; III, Stage III acanthella; IV, Stage IV acanthella; V, Stage V acanthella; ANOM, anomalies.

EXPLANATION OF FIGS. 20-22. The effects of high temperature on the morphogenesis of larval *Moniliformis dubius* in the cockroach intermediate host, *Periplaneta americana*. Each column represents the larvae recovered from one host at the indicated time after infection. The bars having no texture (located at the top of each column) represent the percentage of anomalous larvae in each sample. Absence of such a bar indicates that no anomalous larvae were recovered from that particular cockroach. The textured bar in each column represents the relative composition by larval type of the morphologically normal larvae in each sample. Successive larval types relative to their developmental appearance, if present, are always represented in an ascending order. If the percentage of a recovered larval type was too small to be shown adequately by a textured area of the bar, that particular area of the bar remains untextured and is labeled with the appropriate Roman numeral to the side of the bar. The numbers in parentheses (N) below each indicated time after infection represent the number of larvae recovered from each host.

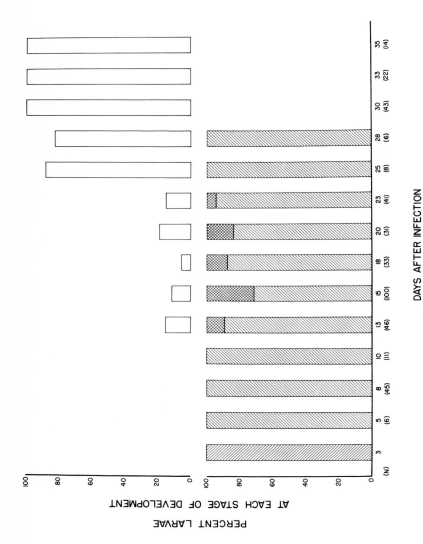

FIG. 20. TEMPERATURE, 35° C; date of infection, 1/9/71; light regime, 12 hours light/12 hours darkness. The larvae represented after about day 13 are clearly delayed in development but appear normal in morphology. After day 28, all larvae recovered were morphological anomalies.

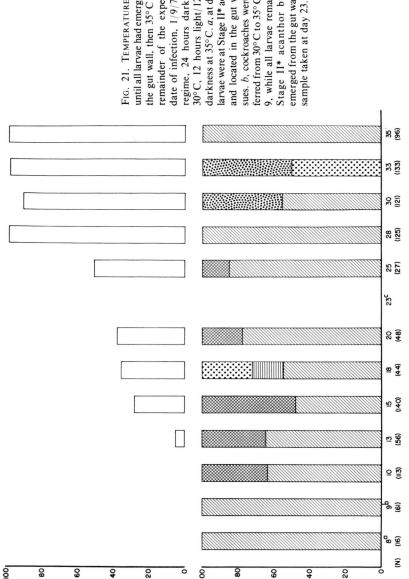

FIG. 21. TEMPERATURE, 30°C until all larvae had emerged from the gut wall, then 35°C for the remainder of the experiment; date of infection, 1/9/71; light regime, 24 hours darkness at 30°C, 12 hours light/12 hours darkness at 35°C. *a*, at day 8 all larvae were at Stage II* acanthor and located in the gut wall tissues. *b*, cockroaches were transferred from 30°C to 35°C at day 9, while all larvae remained at Stage II* acanthor but had emerged from the gut wall. *c*, no sample taken at day 23.

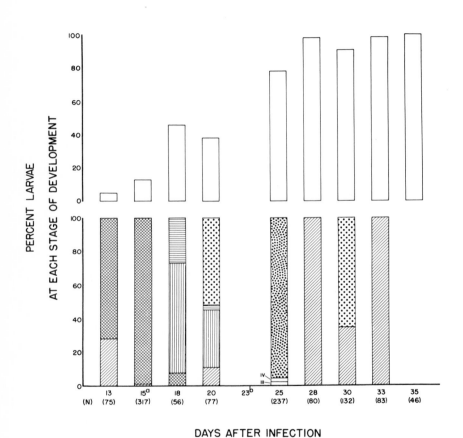

FIG. 22. TEMPERATURE, 30° C until a majority of the larvae had metamorphosed from the acanthor to acanthella stages, then 35° C for the remainder of the experiment; date of infection, 1/9/71; light regime, 24 hours darkness at 30° C, 12 hours light/12 hours darkness at 35° C. *a*, cockroaches were moved from 30° C to 35° C at day 15, when 99.3% of the larvae were at Stage I acanthella. *b*, no sample taken at day 23.

summary effect of 35°C temperature is apparently an increasing rate of larval death and an accelerated host response to the dead larvae.

DISCUSSION

Variability in *M. dubius* infections has been noted by several authors. Moore (1946) suggested that variation in larval size and rate of development might result from variation in hatching, gut penetration, availability of food in a particular hemocoel area, and the rate of nutrient assimilation. King and Robinson (1967) observed that the variation in larval development within a single host was as marked as that in different hosts from the same infection. Quantitative studies by Lackie (1972a) showed that, in light infections, 20-30% of the eggs administered were recovered as cystacanths, and that this percentage diminished as the infection became heavier. In a single host, 0-80% of the eggs might develop successfully. Our infections showed similar variations in the number of developing larvae and their rate of morphogenesis.

It can be seen from tables 1 and 2 and figure 15 that the rates of development we observed at 23°C and 26.5°C were relatively slower than those of Moore (1946) and Lackie (1972a) but compare favorably with that of King and Robinson (1967). Moore observed cystacanths at 23-26°C in 49-55 days while Lackie found terminal development at 27.8°C in 40 days. King and Robinson recorded completion of development at 27°C in 49-58 days. Lackie noted that the slight difference in experimental temperatures did not seem to warrant the compressed time scale he obtained in comparison to that of King and Robinson. He did not mention that his results compare more precisely with the 35-49 days required for morphogenesis in summer, however, for which King and Robinson tentatively suggested seasonal implications.

The studies mentioned previously lead to the apparent conclusion that temperature is the prime controlling factor in the development of *M. dubius* in the intermediate host, but the problem becomes more complex in view of the findings of our series of experiments at 30°C. Maintenance of roaches under continuous darkness in winter, summer, and fall resulted in rates of development that point to a host seasonal effect on the larvae. Another factor to be considered is the influence of photoperiod responses of the host on developing larvae, since there was a significant difference in rates of development at the same season and temperature but under varying light conditions.

The variables influencing morphogenesis are potentially so numerous that systems analysis of the data may be required for a more definitive study. Possible variables affecting the parasite in addition to host rhythms include sex of the host, developmental stage of the host with regard to molting, humidity at which hosts are maintained, and crowding effects in heavy parasite infections. Lackie (1972a) observed that female roaches seem to be prone to larger infections than males, but King and Robinson (1967) saw no such trends, nor did they see any effect of age of host or of crowding. No crowding effects were apparent in the current study.

As the result of a study of the development of *Polymorphus minutus* in the amphipod *Gammarus pulex*, Butterworth (1969) defined three distinct acanthella stages—the early acanthella (corresponding to the Stage I acanthella), the middle acanthella (including the acanthella Stages II-V), and the late acanthella (equivalent to the Stage VI acanthella). Crompton (1970) charted the proportion of time spent in each of these three stages during the development of thirteen acanthocephalans including *Moniliformis dubius*. As extrapolated from Crompton's bar diagram, his estimates of the time spent in each of the acanthella stages by *M. dubius* were as follows: early acanthella, 43%; middle acanthella, 29%; and late acanthella, 28%. These figures require a revision, however, since neither of the publications (Moore, 1946; King and Robinson, 1967) on which Crompton based his calculations clearly documented the duration of the late acanthella stage. From the data in table 6 and the graphic plot of the rate of development at 30°C in figure 15, we have recalculated the relative time spent in each of the three stages by *M. dubius:* early acanthella, 47%; middle acanthella, 37%; and late acanthella, 16%. Thus, more time is spent in the middle acanthella stage and considerably less as a late acanthella than previously estimated. If we included the time spent in the acanthor stages with that spent as an early acanthella (as did Crompton), then the proportion of time spent in this early stage would exceed 50%. Crompton (1970) observed that the proportion of time spent by *P. minutus* in each of the three acanthella stages remained constant even when the total time of larval development changed at different temperatures. With *M. dubius*, the data in tables 1, 2, and 6 and figure 15 suggest that this is not the case. The selective effect of temperature on the duration of the Stage II and IV acanthellas increases the proportion of time spent in the middle acanthella stage as the temperature is lowered from 30°C to 26.5°C and 23°C. The data and analyses presented in this paper suggest that acanthocephalan development is a complex and variable process. Continued accumulation of detailed tabulations of these developmental events will allow us eventually to evolve a stochastic model of acanthocephalan larval morphogenesis using the theory of Markov processes (Bharucha-Reid, 1960).

By far the preponderance of anomalous development occurred at Stage II acanthella, a stage in which a great deal of morphogenesis normally occurs. Voge (1959), in experiments on the sensitivity of developing *Hymenolepis diminuta* to high temperature stress, concluded that the heat-sensitive period in which pathology occurs coincides with the period of maximum larval growth and development. While the anomalies seen in *M. dubius* here cannot be attributed to heat stress alone, since anomalies were present at room temperature also, it is significant that the anomalies occurring both at temperatures allowing normal morphogenesis (23°C-30°C) and at high temperatures (35°C) definitely point to the early stages as a sensitive period particularly subject to developmental aberrations.

The morphogenetic responses of larval *Moniliformis dubius* to temperature closely correspond to the temperature relations of the cockroach intermediate host, *Periplaneta americana*. Adult male *P. americana* provided with food and water *ad libitum* exhibit a temperature preference of 24°C to 33°C (Gunn, 1935). The temperature observed to be the most preferred was 29°C and the cockroaches avoided the higher end of the spectrum more than the lower end. The upper limit of the preferred temperature range is dropped by removing the water—a compensating reaction—and re-established when the water supply is returned. Ramsay (1935) discovered that the outer wax layer of the insect cuticle undergoes a change of phase at 30°C; this modification of the cockroach surface alters water permeability. Thus, water loss from the cockroach host increases radically at about the same temperature (32-33°C) at which normal morphogenesis of the acanthocephalan larvae is disrupted. The host provides a behavioral negative feedback to temperature extremes, thereby maintaining a homeostatic plateau. Overriding the host response to temperature extremes, as we did in the 35°C experiments, results in a runaway or positive feedback in the symbiont leading to "noise" which interferes with those feedback systems normally controlling larval morphogenesis (see Read, 1970, for a discussion of cybernetic principles in symbiosis). The acceptable noise level in the symbiont is exceeded outside of the homeostatic plateau. These events result, at low temperatures, in a negligible rate of development (Lackie, 1972b); at high temperatures, in a loss of developmental regulation or a blockage of morphogenesis; and ultimately—at both ends of the temperature spectrum—in death. In conclusion, the developmental regulation of larval *Moniliformis dubius* has evolved to function best within the limits the host, *Periplaneta americana*, has established as optimal in maintaining its homeostatic plateau.

ACKNOWLEDGMENTS

Our appreciation is due to Prof. Robert Short, who provided the incubators making this study possible and who has given us his friendship and counsel over the years, and to Mr. Ken Saladin, who drew figures 19-22.

REFERENCES CITED

Bharucha-Reid, A.
    1960 Elements of the Theory of Markov Processes and their Applications. New York: McGraw-Hill.

Butterworth, P.
1969 The development of the body wall of *Polymorphus minutus* (Acanthocephala) in its intermediate host *Gammarus pulex.* Parasitology **59**:373-388.

Byram, J. and F. Fisher, Jr.
1973 The absorptive surface of *Moniliformis dubius* (Acanthocephala). II. Functional aspects. Tissue and Cell **5**:553-579.

Crompton, D.
1970 An ecological approach to acanthocephalan physiology. Cambridge Monographs in Experimental Biology **17**:1-125.

Gunn, D.
1935 The temperature and humidity relations of the cockroach. III. A comparison of temperature preference, and rates of desiccation and respiration of *Periplaneta americana, Blatta orientalis* and *Blattella germanica.* Journal of Experimental Biology **12**:185-190.

King, D. and E. Robinson
1967 Aspects of the development of *Moniliformis dubius.* Journal of Parasitology **53**:142-149.

Lackie, J.
1972a The course of infection and growth of *Moniliformis dubius* (Acanthocephala) in the intermediate host *Periplaneta americana.* Parasitology **64**:95-106.

1972b The effect of temperature on the development of *Moniliformis dubius* (Acanthocephala) in the intermediate host, *Periplaneta americana.* Parasitology **65**:371-377.

Moore, D.
1946 Studies on the life history and development of *Moniliformis dubius* Meyer, 1933. Journal of Parasitology **32**:257-271.

Ramsay, J.
1935 The evaporation of water from the cockroach. Journal of Experimental Biology **12**:373-383.

Read, C.
1970 Parasitism and Symbiosis. New York: Ronald Press Co.

Read, C., A. Rothman, and J. Simmons
1963 Studies on membrane transport, with special reference to parasite-host integration. Annals of the New York Academy of Sciences **113**:154-205.

Robinson, E. and A. Jones
1971 *Moniliformis dubius:* X-irradiation and temperature effects on morphogenesis in *Periplaneta americana.* Experimental Parasitology **29**:292-301.

Rotheram, S. and D. Crompton
1972 Observations on the early relationship between *Moniliformis dubius* (Acanthocephala) and the haemocytes of the intermediate host, *Periplaneta americana.* Parasitology **64**:15-21.

Schiller, E.
1959 Experimental studies on morphological variation in the cestode genus *Hymenolepis.* I. Morphology and development of the cysticercoid of *H. nana* in *Tribolium confusum.* Experimental Parasitology **8**:91-118.

Sokal, R. and F. Rohlf
1969 Biometry. The Principles and Practice of Statistics in Biological Research. San Francisco: W. H. Freeman.

Van Cleave, H.
1947 A critical review of terminology for immature stages in acanthocephalan life histories. Journal of Parasitology **33**:118-125.

Voge, M.
1959 Sensitivity of developing *Hymenolepis diminuta* larvae to high temperature stress. Journal of Parasitology **45**:175-181.

THE BIOLOGICAL SIGNIFICANCE OF THE TEGUMENT IN DIGENETIC TREMATODES

by *K. E. Dixon*

ABSTRACT

The structure of the tegument of digenetic trematodes is adapted to serve two primary functions, absorption and protection, and represents a compromise between the demands of the two roles. It is suggested that the covering layer of cytoplasm serves as an absorptive surface for most or even all of the parenchymal cells that are able to establish intermittent connections with it for the purpose of inward and outward transfer of substances. Because the surface must therefore be a delicate structure and hence susceptible to damage, the ability of the parenchymal cells, protected beneath the basal lamina and muscle cells, to regenerate the surface layer of cytoplasm, represents an important protective adaptation. These suggestions extend our understanding of the structural and functional contribution of the surface cytoplasm of the tegument to the organism. They also have implications for our view of the method of formation of the tegument and for the nomenclature of the tegumental structures.

INTRODUCTION

The surface structures of digenetic trematodes have received a great deal of attention, particularly since electron microscopy became more readily available to parasitologists fifteen to twenty years ago. No doubt one of the major reasons for this concentrated study is that the surface can be identified unequivocally, unlike many of the internal structures, which, particularly in the early years, were often difficult to assign with any confidence to an organ system. Notwithstanding these valid technical reasons, recognition of the nutritional contribution of the surface to the parasite was also an important factor. This interest had its genesis in part in and received much of its momentum from the studies of Clark Read and his associates (see Pappas and Read, 1975), directed towards elucidating the functional aspects of molecular trans-

---

K. E. Dixon is Senior Lecturer in Biology at the Flinders University of South Australia.

port across the surface membranes of cestodes and, more recently, trematodes.

The structure of the surface of digenetic trematodes remained enigmatic until the careful electron microscopic studies of Threadgold (1963) clarified its unique syncytial organization and so confirmed and extended the earlier light microscope observations of Hein (1904). Threadgold introduced the term *tegument* to refer to the syncytium, which he defined as consisting of a continuous outer layer of cytoplasm connected to nucleated cell bodies lying beneath the basal lamina and the muscle layers. Lee (1966) has suggested *epidermis* as a simpler alternative but Hockley (1973) has continued to use "tegument" in preference to "epidermis" because he believes the latter does not indicate the unusual structure of the tegument, a view with which I agree (see also Lumsden, 1975). Furthermore, since there is no dermis, "epidermis" is not appropriate. In this article, the term *tegument* will be used as originally defined by Threadgold (1963).

Since Threadgold's description of the tegument of adult *Fasciola hepatica* appeared, many other species have been studied (reviewed in Lee, 1966 and 1972; Hockley, 1973; Lumsden, 1975). There have been few attempts, however, to integrate the structural information with the functional requirements of the parasite. This article takes an ontogenetic approach by examining the functional basis for the cercarial surface structures from which the adult tegument is derived. It is not a review but an attempt to provide an integration of some information already available.

It will be argued that the primary function of the tegument is absorptive and its structure is highly dynamic, in keeping with this function. It is proposed that the surface layer of cytoplasm serves most or even all of the parenchymal cells that are able to make intermittent connections with it. By this means a continuous cytoplasmic pathway is provided throughout the animal. Of equal importance, the structure of the tegument ensures that damage to the delicate absorptive surfaces remains localized and the surface layer of cytoplasm can be regenerated from the parenchymal cells protected beneath the basal lamina and muscle cells.

THE ADULT TEGUMENT

In all adult digenetic trematodes, the tegument consists of two regions (figure 1). At the surface, a layer of cytoplasm rests on a basal lamina, which overlies the major muscle layers. Internal to the basal lamina and the muscle cells, in the parenchyma, lie nucleated cell bodies, which are connected to the surface layer of cytoplasm by narrow tubular extensions passing between blocks of muscle cells and through the basal lamina. Lee (1966 and 1972), Lyons (1973), Hockley (1973), and Lumsden (1975) have reviewed, comprehensively and in detail, the published descriptions of the tegument of adult digenetic trematodes.

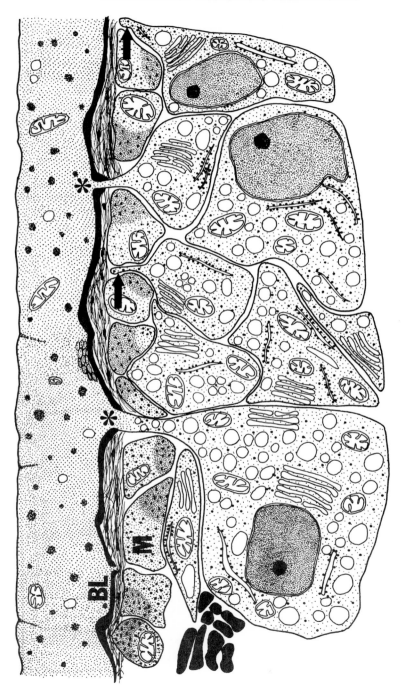

FIG. 1. A DIAGRAMMATIC REPRESENTATION OF THE SURFACE OF A METACERCARIA or young adult of *F. hepatica*. The surface layer of cytoplasm is connected by means of tubular channels (*) to cells lying beneath the muscles (*m*) and the basal lamina (*bl*). A number of processes from other cells (arrows) reach towards the basal lamina and represent incipient channels connecting with the surface layer of cytoplasm.

The adult fluke develops from the cercaria by a process which in some species (e.g., *F. hepatica*, reviewed by Dixon, 1968) is sudden and dramatic enough to be called a metamorphosis. The transition also resembles metamorphosis because it involves in part the replacement of cercarial investments with the tegument of the adult.

THE CERCARIAL TEGUMENT

The exact nature of the cercarial investments and the mechanisms by which they are formed and lost differ in different species. In all species in which early stages of development have been closely studied, the germ ball from which the cercaria develops is surrounded by a thin, nucleated cytoplasmic layer, called a "primitive epithelium" by Dubois (1929) and an "embryonic epithelium" by Dixon and Mercer (1967). I will continue to use the latter term here because "primitive" has little if any meaning in this context, whereas the epithelium is indisputably embryonic (but see Rifkin, 1970; Belton and Belton, 1971). Eventually, a second layer of cells is formed between the embryonic epithelium and the muscle cells and it is from this layer that the cercarial tegument ultimately forms, finally developing into the adult tegument.

In some species, e.g., *Notocotylus attenuatus* (Southgate, 1971) and *Schistosoma mansoni* (Hockley, 1972), the cercarial tegument is formed at an early stage of development of the cercaria, but in other species, e.g., *F. hepatica* (Dixon and Mercer, 1967), it forms during transformation to the metacercarial stage. In other species still, e.g., *Cloacitrema narrabeenensis* (Dixon, 1970; Dixon and Colton, 1975), the cercarial tegument is laid down at intermediate stages of development. No matter when the cercarial tegument appears, however, its formation precedes, usually only briefly, the degeneration and loss of the embryonic epithelium.

Apparently the cercarial tegument can be formed in either of two ways, depending on the species. In the first of these, outer cells of the embryo form a single layer around the developing organism, later fusing into a syncytium, presumably by the breaking down of the lateral membranes. In some species in which the cercarial tegument is formed in this way, the nuclei degenerate completely, as convincingly shown by Matricon-Gondran (1971) in *Cercaria pectinata*, and by Hockley (1972) in *S. mansoni*, and the parenchymal cells eventually establish connections with the surface layer of cytoplasm which remains. In some other species, e.g., *Acanthoparyphium spinulosum* (Bills and Martin, 1966) and *C. narrabeenensis* (Dixon and Colton, 1975), the nuclei apparently do not degenerate but sink down into the sub-tegumental parenchyma, retaining their connection with the surface layer of cytoplasm. The alternative method of formation of the cercarial tegument relies solely on the upflow of cytoplasm from parenchymally situated cells, a process which forms the layer of cytoplasm at the surface, e.g., *F. hepatica* (Dixon, 1968).

Questions as to the functional basis for this structural heterogeneity naturally follow.

THE FUNCTION OF THE CERCARIAL INVESTMENTS

The surface structures of cercariae, both embryonic epithelium and cercarial tegument, have a number of functions to perform: structural, nutritional, and protective, and in formation of the metacercarial cyst wall.

Structural functions. The embryonic epithelium is formed very early in development when the embryo consists of only a few cells. Although similar structures are not formed during embryonic development in the Turbellaria (Skaer, 1973), the formation of a similar envelope during early development of the oncosphere of the cestodes *Eubothrium rugosum* and *Diphyllobothrium latum* has been described by Schaunnsland (1826), cited in Rybicka (1966). One of the functions of this structure is presumably to enclose the cells of the embryo, acting in an analogous way to an egg envelope.

Absorptive functions. Since cercariae grow within the redia or daughter sporocyst, they must acquire the necessary substances for growth. The requirements and mechanisms for the nutrition of intra-redial cercariae have received scant attention. The conclusion that the cercariae must take up substances from the fluid in the redial lumen has been generally accepted although the mechanism of uptake has not been studied. The conclusion seems inescapable, though, that absorption takes place at the cercarial surface, since it is doubtful that the intestinal caeca are functional. In the cercaria of *F. hepatica*, for example, the gut is not patent but consists of a number of isolated cavities in which a secretion of the cells lining the cavities is stored. It can be speculated that this secretion is responsible for digestion of the ventral plug region of the metacercarial cyst wall during excystment, thus permitting the metacercaria to escape from the cyst wall (Dixon, 1966). The gut in this condition could not function as an organ of absorption within the redia, however. Observations of the undeveloped nature of the caecal epithelium of the newly excysted metacercaria are consistent with this reasoning (Bennett, 1975). Køie (1971a), in an ultrastructural study of the caecal epithelium of the intra-redial cercaria of *Neophasis lageniformis*, has also suggested that the caeca in this species are not functional.

On the other hand, positive evidence that the cercarial surface is absorptive is not conclusive. All the descriptions of the embryonic epithelium suggest that it has no specialized organelles or structure to enhance its ability to take up substances from the redial lumen. For example, microvilli or similar devices for increasing the surface area for absorption have not been described. On the contrary, the cytoplasm of the embryonic epithelium appears empty except for a few scattered mitochondria, suggesting that these cells are not metabolically active. It therefore seems likely that while an embryonic epithelium persists, the cercaria has to rely on diffusion of growth substances across the surface, with the diffusion gradient perhaps maintained by active processes at the surface of the inner mass of embryonic cells.

The cells that will ultimately form the cercarial tegument have general ultrastructural characteristics which would enable them to take on an absorptive role although they are not specialized for this function. Some cytochemical studies have demonstrated the presence of enzymes (mainly phosphatases) often associated with absorption, e.g., Dixon (1970) and Dixon, Wetherall, and Colton (1975) in *C. narrabeenensis*, Køie (1971b) in *Zoogonoides viviparus*, and Krupa and Bogitsh (1972) in *S. mansoni*. Direct experimental evidence that the cercariae are capable of absorbing substances is still largely lacking. Dixon (1970) and Dixon et al. (1975) have shown that the cercariae of *C. narrabeenensis* take up horse-radish peroxidase and ^{3}H-glucose. Notwithstanding the paucity of actual data, the logical justification for concluding that the cercarial surface is absorptive is compelling.

The absorptive role of the tegument of the cercaria is ideally suited to animals that have no circulatory system and instead have to rely on diffusion of substances from cell to cell. Gallagher and Threadgold (1967) have suggested that the ramifying parenchymal cells function as a cellular equivalent of a circulatory system. The attractions of this suggestion are enhanced by the recognition that the structure of the tegument provides a continuous cytoplasmic pathway from the exterior to the interior of the body. In this light, the absorptive capacity of the tegument could be utilized more efficiently if it provided a common absorptive surface for most or all of the parenchymal cells. For this to be feasible, the parenchymal cells would have to be able to establish connections with the surface cytoplasm of the tegument, and these cells would then be indistinguishable cytologically from the submuscular cell bodies normally considered as contributing to the tegument. If this view is correct, it further suggests that the tegument is a dynamic and highly active structure, and not a static one as might be inferred from the electron microscope data.

Similar but less explicit suggestions have been made by Lee (1972) and Hockley (1973). Lee has written: "sunken cells are not permanently in contact with the epidermis but . . . they migrate from within the parenchyma to make contact with the epidermis and then perhaps lose contact once their function at the epidermis is complete." Hockley (1973) comments that the subtegumental cells of *S. mansoni* cercariae may be only temporarily connected to the tegument. The view expressed here extends the suggestions of Lee and Hockley further by proposing essentially that many, if not all, of the parenchymal cells establish *intermittent* connections frequently with the surface cytoplasm.

All the observations of metacercarial cyst formation, a process in which the surface structures play an important part (discussed further below), are consistent with the suggestion that many, if not in some species most, of the parenchymal (cystogenic) cells are able to establish connections with the surface at the time of encystment. Furthermore, the failure to observe connections between the surface cytoplasm and the parenchymal cells in *F. hepatica* (Bjorkman and Thorsell, 1964) and in *S. mansoni* cercariae (Hockley, 1972) is evidence in favor of this hypothesis. Additional support comes from reports that the number of connections increases markedly during development of the schistosomulum (Hockley, 1972).

If parenchymal cells are able to establish connections with the cytoplasm intermittently, then an efficient mechanism exists for the regulation of the flow of materials, in both directions, between the interior and the exterior of the animal.

Protective functions. The surface structures of cercariae must also provide protection against the host defenses and the vicissitudes of the external environment. This function is greatly complicated by the fact that the surface is still a major or even the sole organ of absorption and hence is very susceptible to damage. Clegg (1972) has drawn attention to the unsuitability of the surface for protection of the organism. If the surface was composed of a normal cellular epithelium, then damage resulting from injury could only be repaired by the proliferation of remaining undamaged epithelial cells. On the other hand, if the absorptive surface was a tegument, the damage could be restricted to the outer layer of cytoplasm, while the cell bodies remained protected beneath the basal lamina and able to regenerate the surface layer at the appropriate time.

There have been few studies of the regeneration of the tegument in the adult and none at all in the cercaria. Hockley (1972) has reviewed the reports of changes in the tegument of adult flukes following exposure to drugs. No conclusions useful for this article can be reached, however, because damage could be due either to the direct action of the drug on the surface or to more general systemic effects which result in degeneration of the surface.

The tegument is subjected to partial and in some cases complete breakdown during encystment and formation of the metacercarial stage. Therefore, a consideration of the role of the tegument in encystment is relevant.

Encystment functions. The possible evolutionary relationships of metacercarial cyst walls have been recently discussed (Dixon, 1975) on the basis of the degree of protection from the rigors of the environment afforded by cyst walls of different complexities. The postulated sequence (which does not imply a single line of evolution) begins in a simple, single-layered cyst wall, usually composed of carbohydrate-protein complexes (see Dixon, 1975, for examples). An intermediate, more resistant stage has at least two layers of material, formed by the addition of a layer of resistant protein, usually stabilized by disulphide bonds, to the simple cyst wall. The ultimate in protection is reached in those species that encyst in the open, when another layer of protein—a tanned protein—is added and this additional layer is incorporated into an outer cyst wall.

In the formation of the two simpler types of cyst walls, the surface layer of cytoplasm of the tegument has to receive the cystogenous materials from the cells in which they were synthesized, and then secrete them at the appropriate time to the exterior. Transfer of the cystogenous materials is achieved when the parenchymal cells establish connections with the surface cytoplasm. Secretion can be accomplished by a normal apocrine process involving little or no damage to the tegument.

The formation of the most complex cyst walls does, however, involve damage to the surface. In some species, e.g., *C. narrabeenensis* (Dixon and Colton, 1975), the cystogenous precursors of the outer cyst wall are transferred to and secreted from the cercarial tegument. In other species, where apparently the volume of cystogenous material is greater, e.g., *F. hepatica* (Mercer and Dixon, 1967) and *Parorchis acanthus* (Rees, 1967; Cable and Schutte, 1973), the precursor granules of the outer cyst wall are transferred to the embryonic epithelium, which is then shed. In both species, the basal lamina therefore forms, temporarily, the surface of the transforming metacercaria, until upwelling of cytoplasm, principally or perhaps exclusively from the parenchymal cells synthesizing the keratin scrolls, forms a surface layer of cytoplasm. The establishment of a coherent surface is essential for the secretion and unrolling of the individual keratin scrolls (Dixon, 1968).

The regeneration of the metacercarial surface during encystment therefore illustrates the advantages of the tegumental organization for repair of the surface if it is damaged.

CONCLUSIONS

The surface coverings of digenetic trematodes, because they represent the parasite/host or parasite/environment interface, are clearly not evolutionarily conservative structures. On the contrary, they have become adapted to serving a number of functions which must all contribute to the overall economy of the organism.

The preceding analysis implies that the surface layer of cytoplasm of the tegument is primarily adapted as an absorptive organ serving most or all of the parenchymal cells that make more or less frequent but intermittent connections with it. Temporary connections provide a means of regulating the flow of absorbed materials inwards and the flow of secretory materials outwards. Since to fulfill this role the surface is inevitably a delicate structure and therefore susceptible to damage, the ability of the parenchymal cells, protected beneath the basal lamina and muscle cells, to regenerate the surface layer of cytoplasm is clearly an important adaptation. Viewed in this light, arguments as to whether the tegument is formed by the upwelling of cytoplasm from parenchymally situated cells or by the degeneration of nuclei in surface cells and the subsequent establishment of connections with parenchymally situated cells become in part misleading and are largely irrelevant to an understanding of the functions of the tegument. The hypothesis also has implications for the nomenclature of these surfaces. The term *tegument*, as noted earlier, includes both the surface layer of cytoplasm and the nucleated parenchymal cell bodies. If the hypothesis advanced here is correct, however, there is no exclusive association between cell "bodies" (a term which is also no longer appropriate) and the anucleate surface cytoplasm, which should therefore be considered as a separate structure.

Although this analysis has been confined almost exclusively to cercariae and sub-adults, it is obvious that the inferences apply equally to the functions of the tegument in all other stages. The hypothesis is of such a general nature that it can probably be applied profitably to groups other than the Digenea which exhibit this type of structure.

ACKNOWLEDGMENTS

Thanks are due to Dr. J. C. Pearson, Dr. C. P. Read, and Professor J. D. Smyth for their criticism of the ideas on which this article is based.

REFERENCES CITED

Belton, J. C. and C. M. Belton
    1971 Freeze-etch and cytochemical studies of the integument of larval *Acanthatrium oregonense* (Trematoda). Journal of Parasitology **57**:252-260.

Bennett, C. E.
    1975 *Fasciola hepatica*: development of caecal epithelium during migration in the mouse. Experimental Parasitology **37**:426-441.

Bils, R. R. and W. E. Martin
    1966 Fine structure and development of the trematode integument. Transactions of the American Microscopical Society **85**:78-88.

Björkman, N. and W. Thorsell
    1964 On the fine structure and resorptive function of the cuticle of the liver fluke *Fasciola hepatica* L. Experimental Cell Research **33**:319-329.

Cable, R. M. and M. H. Schutte
    1973 Comparative fine structure and origin of the metacercarial cyst in two philophthalmid trematodes, *Parorchis acanthus* (Nicoll, 1906) and *Philophthalmus megalurus* (Cort, 1914). Journal of Parasitology **59**:1031-1040.

Clegg, J. A.
    1972 The schistosome surface in relation to parasitism. *In* Functional Aspects of Parasite Surfaces. A. E. R. Taylor and R. Muller, eds. Oxford: Blackwell. Pp. 23-40.

Dixon, K. E.
    1966 The physiology of excystment of the metacercaria of *Fasciola hepatica* L. Parasitology **56**:431-456.

    1968 Encystment of the cercaria of *Fasciola hepatica*. Wiadomosci Parazytologiczne **14**:689-701.

    1970 Absorption by developing cercariae of *Cloacitrema narrabeenensis* (Philophthalmidae). Journal of Parasitology **56** (Section II):416-417.

    1975 The structure and composition of the cyst wall of the metacercaria of *Cloacitrema narrabeenensis* (Philophthalmidae). International Journal for Parasitology **5**:113-118.

Dixon, K. E. and M. Colton
    1975 The formation of the cyst wall of the metacercaria of *Cloacitrema narrabeenensis*. In preparation.

Dixon, K. E. and E. H. Mercer
    1967 The fine structure of the cystogenic cells of the cercaria of *Fasciola hepatica* L. Zeitschrift für Zellforschung und Mikroskopische Anatomie **77**:331-344.

Dixon, K. E., M. B. Wetherall, and M. Colton
    1975 The body wall of the redia of *Cloacitrema narrabeenensis* (Philophthalmidae) and its role in nutrition of the cercaria. In preparation.

Dubois, G.
    1929 Les cercaires de la region de Neuchâtel. Bulletin de la Société Neuchâteloise des Sciences Naturelles **53**:1-177.

Gallagher, S. S. E. and L. T. Threadgold
    1967 Electron microscope studies of *Fasciola hepatica*. II. The interrelationship of the parenchyma with other organ systems. Parasitology **57**:627-632.

Hein, W.
1904 Zur epithelfrage der Trematoden. Zeitschrift für Wissenschaften Zoologische **77**:400-438.

Hockley, D. J.
1972 *Schistosoma mansoni*: the development of the cercarial tegument. Parasitology **64**:245-252.

1973 Ultrastructure of the tegument of *Schistosoma*. Advances in Parasitology **11**:233-305.

Køie, M.
1971a On the histochemistry and ultrastructure of the redia of *Neophasis lageniformis* (Lebour, 1910) (Trematoda, Acanthocolpidae). Ophelia **9**:113-143.

1971b On the histochemistry and ultrastructure of the tegument and associated structures of the cercaria of *Zoogonoides viviparus* in the first intermediate host. Ophelia **9**:165-206.

Krupa, P. and B. Bogitsh
1972 Ultrastructural phosphohydrolase activities in *Schistosoma mansoni* sporocysts and cercariae. Journal of Parasitology **58**:495-514.

Lee, D. L.
1966 The structure and composition of the helminth cuticle. Advances in Parasitology **4**:187-254.

1972 The structure of the helminth cuticle. Advances in Parasitology **10**:347-379.

Lumsden, R. D.
1975 Surface ultrastructure and cytochemistry of parasite helminths. Experimental Parasitology **37**:267-339.

Lyons, K.
1973 The epidermis and sense organs of the Monogenea and some related groups. Advances in Parasitology **11**:193-232.

Matricon-Gondran, M.
1971 Origine et differenciation du tegument d'un Trematode Digenetique: étude ultrastructurale chez *Cercaria pectinata* (larve de Baccigér baccigér, Fellodistomatides). Zeitschrift für Zellforschung und Mikroskopische Anatomie **120**:488-524.

Mercer, E. H. and K. E. Dixon
1967 The fine structure of the cystogenic cells of the cercaria of *Fasciola*

hepatica L. Zeitschrift für Zellforschung und Mikroskopische Anatomie 77:331-344.

Pappas, P. W. and C. P. Read
1975 Membrane transport in helminth parasites: a review. Experimental Parasitology 37:469-530.

Rees, G.
1967 The histochemistry of the cystogenous gland cells and cyst wall of *Parorchis acanthus* Nicoll, and some details of the morphology and fine structure of the cercaria. Parasitology 57:87-110.

Rifkin, E.
1970 An ultrastructural study of the interaction between the sporocysts and the developing cercaria of *Schistosoma mansoni*. Journal of Parasitology 56 (Section II):284.

Rybicka, K.
1966 Embryogenesis in Cestodes. Advances in Parasitology 4:107-186.

Skaer, R. J.
1973 Planarians. *In* Experimental Embryology of Marine and Freshwater Invertebrates. G. Reverberi, ed. Amsterdam: North-Holland. Pp. 104-125.

Southgate, V. T.
1971 Observations on the fine structure of the cercaria of *Notocotylus attenuatus* and formation of the cyst wall of the metacercaria. Zeitschrift für Zellforschung und Mikroskopische Anatomie 120:420-449.

Threadgold, L. T.
1963 The tegument and associated structures of *Fasciola hepatica*. Quarterly Journal of Microscopical Science 104:505-512.

BIOCHEMICAL AND CYTOCHEMICAL STUDIES OF ALKALINE PHOSPHATASE ACTIVITY IN *SCHISTOSOMA MANSONI*

by Sue Carlisle Ernst

ABSTRACT

Cytochemical localization of alkaline phosphatase activity in *Schistosoma mansoni* showed β-glycerophosphate (βGP), glucose-6-phosphate (G6P), nitrophenylphosphate (NPP), adenosine triphosphate (ATP), and adenosine monophosphate (AMP) all to be hydrolyzed in the invaginations but not on the apical surfaces of the tegument. Of these substrates, only ATP is hydrolyzed in the esophagus or cecum. Biochemical assays indicate that ATP hydrolysis is cysteine-insensitive, totally magnesium-dependent, and partially potassium-dependent, while hydrolysis of the other substrates is totally cysteine-sensitive, only partially magnesium-dependent, and potassium-independent. The pH optimum for all substrates examined is 9.0 or above; however, cytochemical localizations at different pH values and with different techniques gave similar results. The tegumental localization suggests that the invaginations of the tegument represent surface compartments that would facilitate digestive absorptive activity of this membrane. Furthermore, the localization of nonspecific alkaline phosphatase activity in the tegument, but not in the esophagus or cecum, may reflect regional differences in function.

INTRODUCTION

Many studies have shown that *Schistosoma mansoni* adults acquire nutriments through carrier-mediated mechanisms, and there is some evidence that both the tegument and the cecum are involved in this process. In other helminths, such absorptive functions frequently are associated with a digestive process in such a way as to provide a kinetic advantage for the absorption of

---

Sue Carlisle Ernst is Assistant Professor of Anatomy at Temple University School of Medicine.

products of the digestive activity (see Pappas and Read, 1975, for review). One such enzyme (or enzyme complex) frequently associated with absorptive surfaces is alkaline phosphatase, and its activity in *S. mansoni* has been the subject of numerous previous biochemical and cytochemical investigations (e.g., Cesari, 1974; Bogitsh and Krupa, 1971). The present study further examines the hydrolysis of phosphate esters by the tegument, esophagus, and cecum, which are the potential digestive absorptive surfaces of *S. mansoni*.

MATERIALS AND METHODS

General

Worms were removed from mice with patent infections of *Schistosoma mansoni* by perfusion of the vaculature with citrated saline. They were rinsed in saline without citrate before use.

Cytochemical

Worms were fixed at room temperature for 30-45 minutes in either 6% glutaraldehyde, 0.25% glutaraldehyde + 1% paraformaldehyde, or 4% paraformaldehyde in .1M cacodylate buffer, pH 7.5. They were then rinsed in .1M tris-maleate (TM) or tris-HCl (T-HCl) buffer at pH 7.5 or 9.0 and frozen in the same buffer with .25M sucrose added. 50μ sections were cut on an International Cryostat. In some cases non-frozen tissue was sectioned using a Smith-Farquhar tissue slicer set at 25μ. In either case the sections were collected in buffer and incubated in one of several media for 15-45 minutes at room temperature. For incubation at pH 7.5, the incubation medium was composed of 100 mM TM, 10 mM substrate, 5 mM magnesium chloride, 40 mM potassium chloride, and 2.0 mM lead nitrate. The substrates used were β-glycerophosphate (βGP), nitrophenylphosphate (NPP), and glucose-6-phosphate (G6P). Controls consisted of deletion of the substrate or addition of 20 mM cysteine to the medium. Localizations also were carried out using the procedure of Wachstein-Meisel (1957) with adenosine triphosphate (ATP), adenosine monophosphate (AMP), NPP, or G6P as substrate. Controls consisted of deletion of magnesium from the media. Incubations were run at pH 9.0 using a modified Gomori (1952) calcium-cobalt technique. The incubation medium contained 10 mM substrate (β6P, NPP, or G6P), 0.2M 2-amino-2-methyl-1,3 propandiol buffer, 40 mM potassium chloride, 10 mM magnesium chloride, and 0.1 M calcium chloride. Sections were rinsed in 2% lead nitrate (Wetzel et al., 1967) or in 2% cobalt acetate followed by 1% ammonium sulfide (see Pearse, 1960). Controls consisted of deletion of the substrate or addition of 20 mM cysteine. Localizations also were carried out using the procedure of S. A. Ernst (1972), except that 10 mM βGP, NPP, and G6P were all used as substrate. Addition of 20 mM cysteine or ouabain to the media and deletion of the potassium served as controls.

Following the incubation, the sections were rinsed in .1 M cacodylate buffer at pH 7.5, osmicated for 30 minutes in 1% osmium tetroxide in .1 M cacodylate, dehydrated in a graded alcohol series and embedded in the low viscosity medium of Spurr (1969). Sections were examined both stained and unstained in a Philips 300 electron microscope.

Biochemical

Worms were homogenized in distilled water using a mechanically driven ground glass homogenizer. The homogenates were used immediately. Homogenate was added to the incubation medium to give a final concentration of 100 mM TM or T-HCl buffer at the appropriate pH, 10 mM substrate (βGP, NPP, G6P, ATP, AMP), 5 mM magnesium chloride, and 40 mM potassium chloride.

Substrate concentration curves showed 10 mM to be near optimum for all substrates except ATP, which was above 25 mM. All further experiments were carried out using 10 mM substrate concentrations. The medium was modified by the deletion of one or both cations or by the addition of 20 mM cysteine. Reactions were run for 30 minutes at 37° C and stopped with trichloroacetic acid (final concentration 5%). Phosphate was determined according to the technique of Chen et al. (1956). Nitrophenol determinations were performed as previously described (S. C. Ernst, 1975). All samples were run in quadruplicate and measured against an appropriate blank.

RESULTS

The pH optimum for the hydrolysis of all substrates examined (βGP, NPP, G6P, ATP, AMP) is 9.0 or above and hydrolysis of these substrates

---

FIGURES 1-5 OVERLEAF

FIGS. 1-3. LOCALIZATION OF ALKALINE PHOSPHATE ACTIVITY AT pH 7.5. Lead phosphate reaction product (R) is restricted to the invaginations of the plasmalemma and does not occur on the apical membrane (M). The reaction product frequently occurs in flask shaped pockets (open triangles) which appear to open into the main channels. Neither the rod shaped granules (circle) nor the multilaminate granules (arrows) contain reaction product, with the exception of a few multilaminate granules (asterisk) which appear to contain a very small amount. 1. Substrate βGP. Counterstained with lead citrate and uranyl acetate. ×16,400. 2. Higher magnification of area similar to that seen in figure 1 with no counterstain. ×26,800. 3. Reaction sites are identical to those seen in figures 1 and 2, in paraformaldehyde (4%) fixed tissue with NPP as substrate. No counterstain. ×39,600.

FIGS. 4-5. TISSUE REACTED WITH βGP AT pH 7.5 AND COUNTERSTAINED. 4. Reaction product (arrows) occurs on the limiting membranes of the parenchymal cells and of the subtegumental cell bodies (CB). The lead staining of the Golgi apparatus (G) is nonenzymatically produced. ×22,700. 5. Cecal cytoplasm showing no activity in the lumen (L) or in the cytoplasm. Only a small amount of reaction product is associated with the basal infoldings (arrow). Golgi (G) staining is nonenzymatically produced. ×19,800.

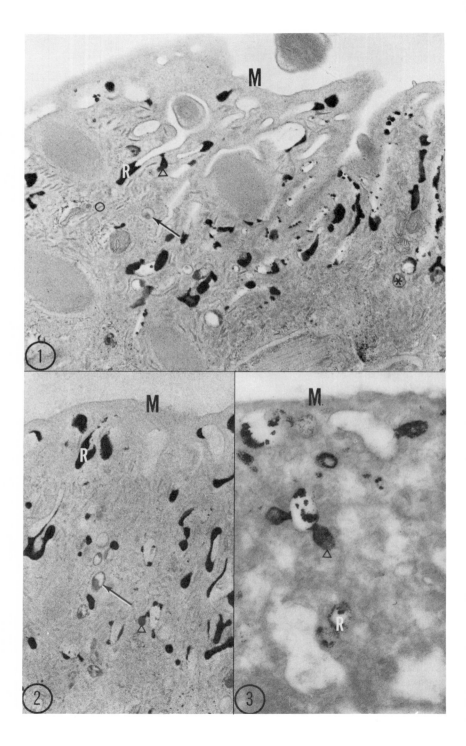

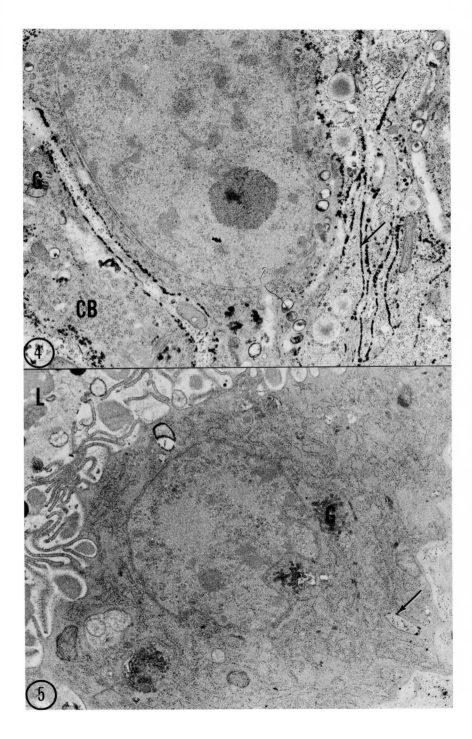

is greatly reduced at pH 7.0-7.5. Cytochemical localization of alkaline phosphatase activity was carried out with all substrates except ATP and AMP at both pH 7.5 and 9.0. Biochemical measurements of the effect of various fixatives showed much of the activity to be destroyed by fixation in 6% glutaraldehyde, while fixation in either 4% paraformaldehyde or 1% paraformaldehyde + .25% glutaraldehyde preserved as much as 75-90% of the enzyme activity. No differences in sites of reaction product occur at the different pH values or with different fixatives. In all cases the substrates are hydrolyzed by the tegument; the reaction product appears in the infolded channels, however, and not on the apical portion of the plasmalemma (figures 1-3). Frequently, the reaction product appears in flask-shaped secondary invaginations off the channels (figures 1-3). Such flask-shaped invaginations were previously noted by Morris and Threadgold (1968). The multilaminate and rod-shaped granules generally do not react (figures 1, 2, and 6); however, a small amount of staining sometimes occurs in a few vesicles that resemble multilaminate granules (figures 1 and 6). It is possible that these reacted granules have fused with the plasmalemma and have secondarily obtained the activity. Profiles such as that seen in figure 6 suggest this hypothesis. Other authors (Bogitsh and Krupa, 1971; Halton, 1967; Nimmo-Smith and Standen, 1963) using different substrates and techniques, have reported that the dorsal tegument of the male reacts more strongly than the ventral tegument, while the female stains uniformly. This differential staining has been observed in the present study as well. Reaction product also appears on the limiting membranes of the parenchymal cells, subtegumental, and subesophageal cell bodies (figures 4 and 7).

No activity occurs in the esophagus with βGP, NPP, or G6P as substrate. Furthermore, no activity is associated with the cecal cytoplasm, with the exception of some reaction product on the basal infoldings (figures 5 and 7). At pH 7.5, lead binds to the Golgi membranes (figures 4 and 5). This non-enzymatically produced staining, which occurs both in the presence of cysteine and in the absence of substrate at pH 7.5, does not occur at pH 9.0 (figure 7). All of the remainder of the staining at both pH values is inhibited either by addition of cysteine to the medium or by deletion of the substrate. These observations at pH 9.0 are consistent with biochemical data which

---

FIGS. 6-7. LOCALIZATION OF ALKALINE PHOSPHATASE ACTIVITY WITH CALCIUM-COBALT METHOD AT pH 9.0. No counterstain. 6. Tegumentary localization is identical to that seen in figures 1-3, in which the reaction product (R) is limited to the invagination and does not occur on the apical portion of the membrane (M). Multilaminate granules (arrow) are usually unreacted; however, some which may be fused with external membrane may have acquired reaction product secondarily (asterisk). Substrate βGP. ×27,400. 7. Results are identical to those obtained with other techniques (compare figures 4 and 5) in that the cecal lumen (L) and cytoplasm (C) contain no reaction product. Small amounts occur in the basal infoldings (arrow). The limiting membranes of the parenchymal cells are reacted (R). Substrate NPP. ×14,600.

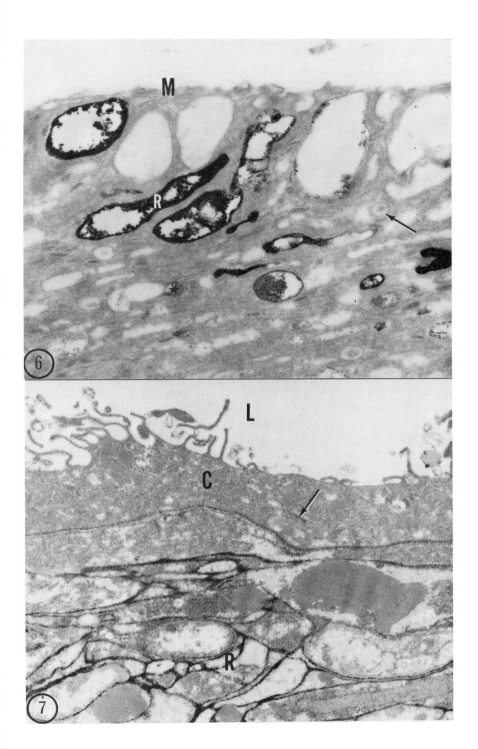

show the hydrolysis of βGP, NPP, G6P, and AMP to be totally inhibited by cysteine (inhibitor : substrate = 2:1). It is not possible to demonstrate total cysteine inhibition biochemically at pH 7.5, although the cytochemical reaction is totally inhibited by cysteine at pH 7.5 as well.

Use of ATP, AMP, G6P, or NPP as substrate with a Wachstein-Meisel procedure produced tegumental and parenchymal staining identical to that seen with the other procedures (figure 8). ATP also is hydrolyzed in the esophagus, however, and to some extent in the cecal lumen. The reaction in the cecum is slight and is associated with the luminal side of the apical folds and with the amorphous material commonly found in the lumen (figure 9). In contrast to results obtained with the other substrates, the luminal side of the lining of the anterior esophagus hydrolyzed ATP (figure 10). In the posterior esophagus, the reaction product found in the lumen is associated with luminal contents rather than the apical membrane. Reaction product is also found in the basal infoldings of the esophageal folds (figure 11). The hydrolysis of ATP is totally dependent on the presence of magnesium; and, as seen in figure 12, deletion of magnesium from the medium results in inhibition of the formation of the reaction product. Only a few small scattered deposits of lead are seen. This is consistent with the biochemical data, which show that deletion of magnesium completely inhibits the hydrolysis of ATP. Chemical measurements of the hydrolysis of the other substrates show them to be only partially inhibited (20-40%) by the deletion of magnesium from the medium. The hydrolysis of ATP also differs from that of the other substrates, in that it is 25% inhibited by deletion of potassium from the medium and it is not affected by cysteine.

Use of the technique developed by S. A. Ernst (1972) for the localization of transport ATPase produces results similar to those obtained with NPP, G6P, and βGP with other techniques at pH 9.0. No cysteine-insensitive or ouabain- and K-sensitive activity corresponding to that localized with ATP as substrate at pH 7.5 could be demonstrated after use of standard concentrations of components. Further modifications of the reaction mixture were not attempted.

FIGS. 8-12. LOCALIZATION OF ALKALINE PHOSPHATASE ACTIVITY WITH THE WACHSTEIN-MEISEL PROCEDURE AND ATP AS SUBSTRATE. No counterstain. 8. Tegument shows localization identical to that obtained with other techniques and substrates. Reaction product (R) is in invaginations and not on the apical membrane (M). Some reaction product is associated with the basal infoldings (B). ×22,200. 9. No reaction product is seen in the cecal cytoplasm (C). A small amount is associated with the folds and luminal contents (arrows). ×19,600. 10. Anterior esophagus with reaction product associated with the luminal side of the folds (arrow) and with the basal membrane (B). ×9,200. Inset: The reaction product (arrow) is restricted to the luminal surface of the membrane. ×25,400. 11-12. In the posterior esophagus the reaction product (double arrow) is associated with luminal inclusions rather than the membrane. The basal infoldings (single arrow) are heavily reacted. ×19,800. 12. Deletion of magnesium from the medium abolishes the reaction in the basal infoldings (arrows) as well as in the lumen. Only a few small scattered lead deposits occur. ×19,800.

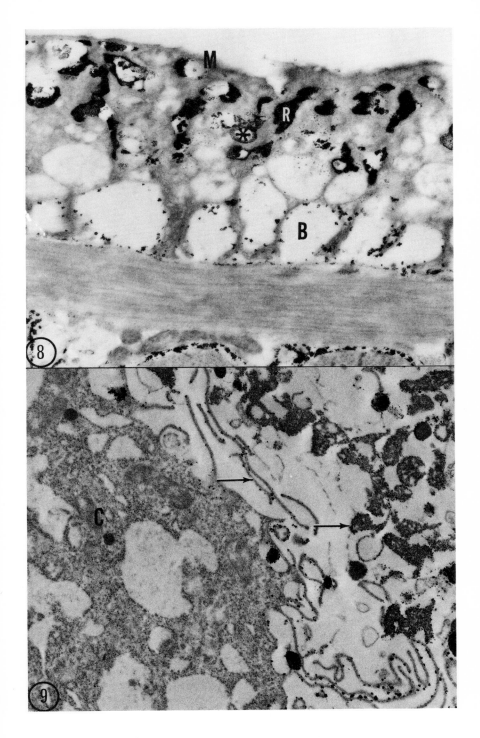

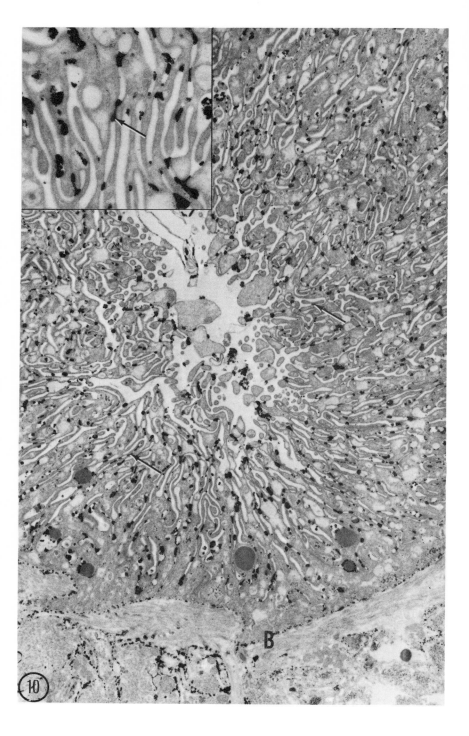

10   B

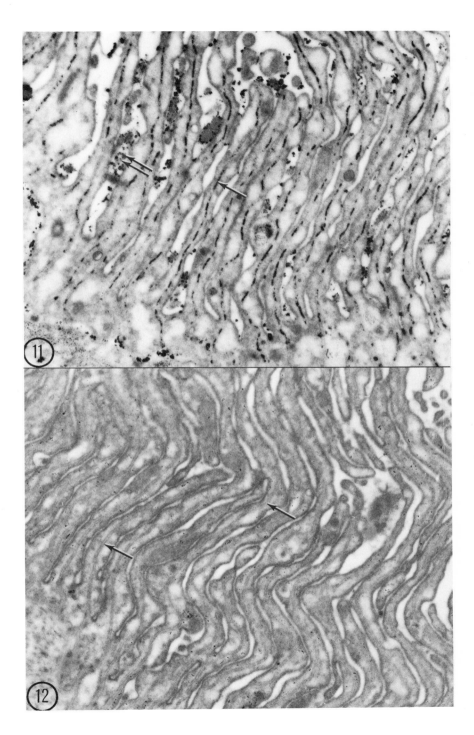

DISCUSSION

Previous ultrastructural cytochemical studies have reported that the tegumental invaginations are the site of hydrolysis of a variety of phosphate esters including βGP, thiamine pyrophosphate, and nucleoside mono-, di-, and triphosphates (Morris and Threadgold, 1968; Bogitsh and Krupa, 1971), and the present study confirms these observations by use of different techniques and extends them to include other substrates. The results of these previous studies, as well as the present study, clearly demonstrate that these invaginations represent specialized areas of the plasmalemma in that they are enzymatically different from the apical portion of the membrane with which they are continuous. The physical configuration of these invaginations and the presence of an intrinsic digestive enzyme in them suggest that they serve to compartmentalize digestive absorptive activity in a manner analogous to that described for the cestode tegument (Dike and Read, 1971a, b) and the acanthocephalan surface (Uglem et al., 1973). The recent studies of Levy and Read (see Pappas and Read, 1975, for review) support this conclusion. It is also possible that in addition to enhancing digestive absorptive activity, these compartments could serve to regulate a microenvironment suitable for optimal function of the enzymes and transport proteins located in them.

The esophageal lining is a modified portion of the tegument and continuous with it (Dike, 1971). However, the surface area of this region is greatly amplified and resembles other epithelia specialized for absorption. It was, therefore, of interest to examine this surface for alkaline phosphatase activity. Of the substrates tested, only ATP is hydrolyzed in the esophageal region. In the anterior portion the enzyme is associated with the luminal side of the plasmalemma, while in the posterior portion the activity is found in the basal infoldings. Since acid phosphatase activity had also been found in this location (Bogitsh and Shannon, 1971; S. C. Ernst, 1975), the possibility existed that the activity observed at pH 7.5 was merely residual acid phosphatase activity. The fact that the ATP hydrolysis is totally magnesium-dependent, whereas the acid phosphatase activity is not, does not support this conclusion. Furthermore, both βGP and NPP are hydrolyzed at this site at acid pH while neither of these substrates is hydrolyzed in the esophagus at pH 7.5 or 9.0.

The hydrolysis of ATP differs from that of the other substrates tested, in that the former requires the presence of magnesium, involves a potassium-sensitive component, and is insensitive to the presence of cysteine. These characteristics are common to Mg^+- and Na^+-K^+ATPase complexes found in other systems, and it is probable that the activity localized using the Wachstein-Meisel (1957) procedure represents Mg^+ ATPase. Use of the S. A. Ernst (1972) procedure for localization of Na^+K^+ATPase did not differentiate sites of activity that were cysteine-insensitive. This failure probably results from use of suboptimal concentrations of the various components of the incubation medium for the visualization of a small portion of the total ATPase activity. The complete characterization of the optimal kinetic parameters of Na$^+$K$^+$ATPase in *S. mansoni* was beyond the scope and purpose of this study.

Biochemical studies (Conde-del Pino et al., 1968; Nimmo-Smith and Standen, 1963; Smithers et al., 1965; Cesari, 1974) of alkaline phosphatases in *S. mansoni* indicate the presence of multiple enzymes active at alkaline pH, and observation of reciprocal inhibition of substrates (Ernst, unpublished data) supports this conclusion. It is, therefore, noteworthy that with the exception of ATP, no differences in sites of hydrolysis were resolved with different substrates or different techniques. This implies either that multiple enzymes occupy the same cytological sites or that some of these enzymes are exceedingly sensitive even to mild fixation with paraformaldehyde. A third possibility is that the substrate-specific portions of the activity represent such a relatively small amount of the total activity that the cytochemical procedures employed are not sensitive enough to differentiate such activity from a non-specific enzyme hydrolysis. The resolution of the question of different cytological sites for these multiple enzymes will require purification of the specific enzyme fractions, kinetic characterization of the various enzymes, and careful application of cytochemical procedures.

It is interesting that the cytochemical localization of surface-associated non-specific alkaline and acid phosphatase activities shows regional localization, with the alkaline phosphatase activity being on the tegument but not in the cecum portion of the esophagus, whereas the acid phosphatase activity is located in the posterior esophagus and the cecum (S. C. Ernst, 1975). The anterior esophagus appears to be a "buffer zone" possessing neither enzyme. It might be speculated that this regional localization could be the reflection of a mechanism for regulation of pH by the worm. If this were the case, it would have a great effect on the functioning of carrier-mediated absorption of nutrients by these surfaces.

ACKNOWLEDGMENTS

Research sponsored by the U. S. Army Medical Research and Development Command under Contract No. DADA-17-72C-2156. Infected mice were provided by the U. S.-Japan Cooperative Medical Science Program-NIAID.

I wish to thank Ms. Laura M. May for her excellent technical assistance during the course of this study and preparation of the manuscript.

REFERENCES CITED

Bogitsh, B. J. and P. L. Krupa
    1971 *Schistosoma mansoni* and *Haematoloechus medioplexus*:

Nucleosidediphosphatase localization in tegument. Experimental Parasitology **30**:418-425.

Bogitsh, B. J. and W. A. Shannon, Jr.
1971 Cytochemical and biochemical observations on the digestive tracts of digenetic trematodes. VIII. Acid phosphatase activity in *Schistosoma mansoni* and *Schistosomatium douthitti.* Experimental Parasitology **29**:337-347.

Cesari, I. H.
1974 *Schistosoma mansoni*: Distribution and characteristics of alkaline and acid phosphatase. Experimental Parasitology **36**:405-415.

Chen, P. S., Jr., T. Y. Toribara, and H. Warner
1956 Microdetermination of phosphorus. Analytical Chemistry **28**:1756-1758.

Conde-del Pino, E., A. M. Annexy-Martínez, M. Pérez-Vilar, and A. A. Cintrón-Rovera
1968 Studies in *Schistosoma mansoni.* II. Isoenzyme patterns for alkaline phosphatase, isocitric dehydrogenase, glutamic oxalacetic transaminase and glucose-6-phosphate dehydrogenase of adult worms and cercariae. Experimental Parasitology **22**:288-294.

Dike, S. C.
1971 Ultrastructure of the esophageal region in *Schistosoma mansoni.* American Journal of Tropical Medicine and Hygiene **20**:552-568.

Dike, S. C. and C. P. Read
1971a Tegumentary phosphohydrolases of *Hymenolepis diminuta.* Journal of Parasitology **57**:81-87.

1971b Relation of tegumentary phosphohydrolase and sugar transport in *Hymenolepis diminuta.* Journal of Parasitology **57**:1251-1255.

Ernst, S. A.
1972 Transport adenosine triphosphatase cytochemistry. II. Cytochemical localization of ouabain-sensitive, potassium dependent phosphatase activity in the secretory epithelium of the avian salt gland. The Journal of Histochemistry and Cytochemistry **20**:23-38.

Ernst, S. C.
1975 Biochemical and cytochemical studies of esophagus, cecum, and tegument in *Schistosoma mansoni*: Acid phosphatase and tracer studies. Journal of Parasitology **61**:633-647.

Gomori, G.
1952 Microscope Histochemistry. Chicago: University of Chicago Press.

Halton, D. W.
    1967 Studies on phosphatase activity in Trematoda. Journal of Parasitology **53**:46-54.

Morris, G. P. and L. T. Threadgold
    1968 Ultrastructure of the tegument of adult *Schistosoma mansoni*. Journal of Parasitology **54**:15-27.

Nimmo-Smith, R. H. and O. D. Standen
    1963 Phosphomonoesterases of *Schistosoma mansoni*. Experimental Parasitology **13**:305-322.

Pappas, P. W. and C. P. Read
    1975 Membrane transport in helminth parasites: A review. Experimental Parasitology **37**:469-530.

Pearse, A. G. E.
    1960 Histochemistry: Theoretical and Applied, 2nd ed. Boston: Little, Brown, Co.

Smithers, S. R., D. B. Roodyn, and R. J. M. Wilson
    1965 Biochemical and morphological characteristics of subcellular fractions of male *Schistosoma mansoni*. Experimental Parasitology **16**:195-206.

Spurr, A. R.
    1969 A low-viscosity expoxy resin embedding medium for electron microscopy. Journal of Ultrastructural Research **26**:31-43.

Uglem, G. L., P. W. Pappas, and C. P. Read
    1973 Surface aminopeptidase in *Moniliformis dubius* and its relation to amino acid uptake. Parasitology **67**:185-195.

Wachstein, M. and E. Meisel
    1957 A comparative study of enzymatic staining reactions in the rat kidney with necrobiosis induced by ischemia and nephrotoxic agents (mercuhydrin and DL-serine). The Journal of Histochemistry and Cytochemistry **5**:204-220.

Wetzel, B. K., S. S. Spicer, and R. G. Horn
    1967 Fine structural localization of acid and alkaline phosphatases in cells of rabbit blood and bone marrow. The Journal of Histochemistry and Cytochemistry **15**:311-334.

PROLINE IN FASCIOLIASIS: II. CHARACTERISTICS OF PARTIALLY PURIFIED ORNITHINE - δ - TRANSAMINASE FROM *FASCIOLA*

by *Jeanette C. Ertel and Hadar Isseroff*

ABSTRACT

Ornithine-δ-transaminase from *Fasciola* was partially purified by differential centrifugation and precipitation by ammonium sulfate. The K_m's for the purified transaminase were 2.33 mM and 2.24 mM for ornithine and α-ketoglutarate, respectively. The pH optimum (8.0) and the inhibitory effect of other amino acids is similar to that of the mammalian enzyme. Pyridoxal dependence and inhibition by pyridoxal inhibitors were also demonstrated. The only observed peculiarity of the fluke enzyme that might explain its high activity appears to be a high tolerance for α-ketoglutarate.

INTRODUCTION

Recent studies have shown that free proline occurs in high levels in *Fasciola hepatica* (Kurelec and Rijavec, 1966) and in the bile of animals infected with this trematode (Isseroff et al., 1972). Ornithine-δ-transaminase [E.C. 2.6.13 L-ornithine: 2-oxoacid aminotransferase], hereafter called OTA, catalyzes the formation of the immediate precursor to proline, Δ^1-pyrroline-5-carboxylic acid, from ornithine and α-ketoglutarate. In a previous paper (Ertel and Isseroff, 1974) this enzyme was shown to occur in homogenates of *Fasciola* and to have seven and ten times the activity of the OTA in rat and rabbit liver, respectively. It was the high activity of the OTA, coupled with the lack of an enzyme capable of breaking down proline, which explained the high levels of proline in the worm. These data also suggested that the excessive proline in the biliary fluid of infected animals originated in the worm. The present study deals with some of the properties of this enzyme in partially purified preparations from *Fasciola*.

Jeanette Ertel is a Research Associate at SUNY College at Buffalo. Hadar Isseroff is Associate Professor of Biology at SUNY College at Buffalo.

METHODS AND MATERIALS

Preparation of ornithine-δ-transaminase enzyme

Infections of *Fasciola hepatica* were maintained in New Zealand rabbits and Wistar rats according to the methods of Isseroff and Read (1969). Rabbits or rats with 75- to 90-day infections were killed to obtain the worms. After removal from the bile ducts, they were rinsed in ice cold 0.05M tris-chloride buffer, pH 8.0, blotted on filter paper, and weighed. For homogenization, buffer was added in the proportion of 10 ml/g of fluke tissue, and the process was carried out at 0°C for 5 minutes in a Potter-Elvehjem homogenizer. The following procedure was carried out at 0°C. The homogenate was centrifuged at 12,000 × *g* for 20 minutes and the supernatant was removed. Sufficient ammonium sulfate (Fisher Chem. Co.) was then added to achieve 60% saturation. After standing for 30 minutes, the resulting suspension was centrifuged at 3000 × *g* for 10 additional minutes before being recentrifuged at 3000 × *g* for 10 minutes. The partially isolated enzyme was dissolved in a small amount of 0.05M tris-chloride buffer, pH 8.0. This was dialyzed in the cold against two changes of water for 2 hours and subsequently lyophilized.

Enzyme Assays

Δ^1-pyrroline-5-carboxylic acid (PCA), the product of the OTA reaction, was assayed by Strecker's method as in a previous study (Ertel and Isseroff, 1974) but with the following modification: the concentration of α-ketoglutarate was increased to 20 mM (except where indicated) because this concentration was found to be saturating and non-inhibitory to the reaction. Enzyme assays were carried out with three or more samples per experiment. We used reagents of the highest purity available, purchased from the Sigma Chemical Company unless another source is indicated. Specific activity, or V_{max}, is defined as μmoles of product/mg of protein per hour. Protein was determined by the method of Lowry et al. (1951) with bovine serum albumin as a standard. In determining K_m's the concentration of protein was approximately 0.01 mg/ml of incubation mixture. This concentration was found to be well within the linear range when velocity was measured as a function of protein concentration using 35 mM ornithine and 20 mM α-ketoglutarate. Time-course studies showed that the reaction was linear for at least 30 minutes with these concentrations of enzyme and substrates.

Determination of α-ketoglutarate

For determination of α-ketoglutarate levels in *Fasciola*, the worms were removed from the bile ducts of infected rats killed by etherization, then were quickly blotted on filter paper and weighed. A number of flukes giving an approximate wet weight of 0.6g were immediately placed in 4.0 ml of 0.6M perchloric acid and homogenized in a Potter-Elvehjem homogenizer. One ml of acid was used to wash out the tube and the precipitated protein was removed by centrifugation. Perchloric acid was then removed from the supernatant by adding 1.0M K_2CO_3 and the extract adjusted to pH 7.2. Liver samples were obtained from healthy rats dispatched with ether and treated similarly except that 0.5 to 1.0 g wet weight of tissue was used. The concentration of α-ketoglutarate was then tested by an enzymatic assay using glutamic dehydrogenase (Bergmeyer and Bernt, 1965).

RESULTS

The specific activity of the 60-75% fraction was four to five times that of the crude homogenate. Table 1 shows a typical purification of OTA from two grams wet weight of fluke material, with an approximate yield of 20%.

Previous studies in our laboratory have shown the pH optimum for OTA in crude homogenates of *Fasciola*, rat liver, and rabbit liver to be 8.0. A similar pH optimum was previously reported for rat liver (Katunuma et al., 1964) and pig kidney (Jenkins and Tsai, 1970). In the present study with 0.025M tris-chloride buffer, the purified enzyme also showed maximal activity at pH 8.0 (figure 1). All subsequent experiments were, therefore, carried out in 0.025M tris-chloride buffer, pH 8.0.

Figure 2 shows the effect of temperature on OTA. In the presence of ornithine, α-ketoglutarate, and pyridoxal phosphate, the optimum temperature for 30-minute incubations was found to be 45°C, with a sharp drop in activity above 50°C.

The Michaelis constants were determined for each of the substrates, ornithine and α-ketoglutarate, according to the methods of Lineweaver and Burk (1934). With preparations having specific activities of about 19-26 μmoles of PCA/mg of protein per hour, the K_m values for ornithine and α-ketoglutarate were determined to be 2.33 and 2.24 mM, respectively (table 2 and figures 3 and 4).

TABLE 1

COMPARISON OF ACTIVITIES AND YIELDS OF OTA IN THE
CRUDE HOMOGENATE AND AFTER PARTIAL PURIFICATION

Fraction	Total units	Specific activity	Purification	% Yield
crude homogenate	661	5.8	1	100
ammonium sulfate fraction (60-75%)	133	25.0	4.3	20

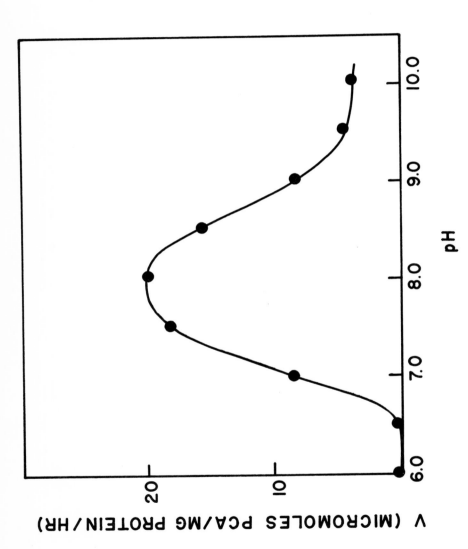

FIG. 1. The effect of pH on the activity of OTA, with 0.025M Tris Cl buffer. Each point is the mean of three samples.

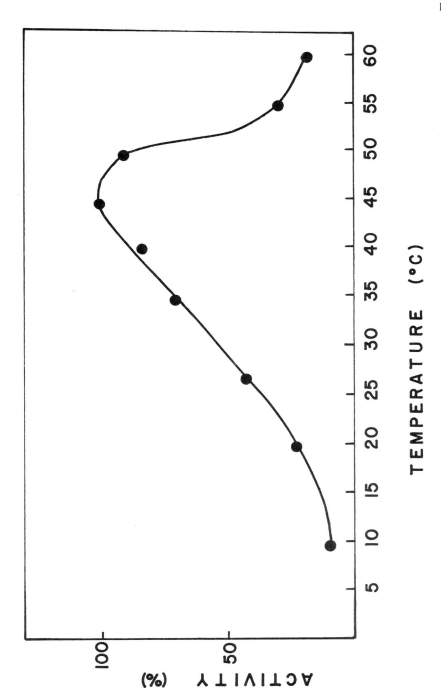

FIG. 2. The effect of temperature on OTA activity in 30 minute incubations. Each point is the mean of five samples.

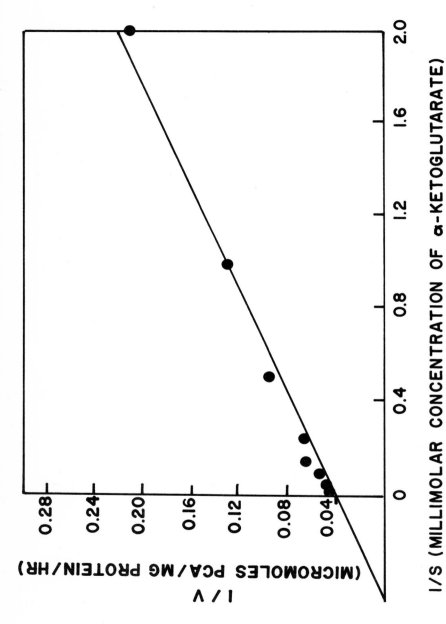

Fig. 3. Typical Lineweaver-Burk plot of ornithine concentration versus velocity of OTA reaction. Each point is the mean of three samples.

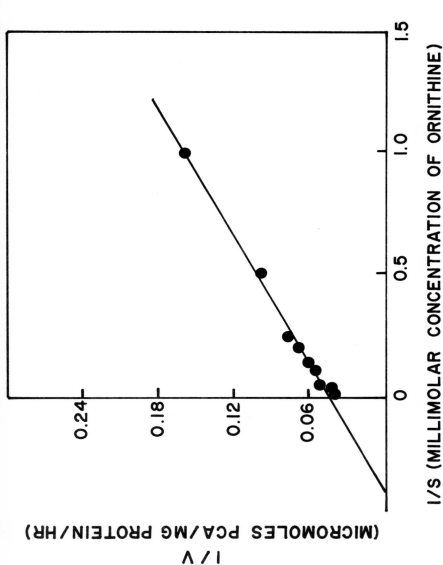

FIG. 4. Typical Lineweaver-Burk plot of α-ketoglutarate concentration versus velocity of OTA reaction. Each point is the mean of three samples.

TABLE 2

K_m AND V_{max} VALUES FOR OTA

K_m values are in μmoles of ornithine or α-ketoglutarate per ml. V_{max} values are in μmoles of PCA produced/mg of protein per hour. Ornithine was present at 35 mM for K_m determinations of α-ketoglutarate. α-ketoglutarate was present at 10 mM for K_m determinations of ornithine. Values shown are the means of three experiments.

Substrate	$K_m \pm$ S. E.	$V_{max} \pm$ S. E.
ornithine	2.33 ± 0.229	18.7 ± 4.001
α-ketoglutarate	2.24 ± 0.198	26.2 ± 1.217

TABLE 3

EFFECT OF PYRIDOXAL-5-PHOSPHATE ON THE ACTIVITY OF ORNITHINE-δ-TRANSAMINASE

Ornithine and α-ketoglutarate were present at 35.0 and 20.0 μmoles/ml, respectively. The pH was 8.0 in all vessels. Each value is the mean of three determinations. The effect given is significant at $P < 0.01$ in a two tailed Student's t test.

Addition of pyridoxal PO_4	μmoles of PCA produced/mg of protein/hour $\pm$ S.E.	Significant effect (%)
none	8.79 ± 0.123	—
2.5 μmoles/ml	17.93 ± 0.520	$+104$

Although the crude homogenate did not require added pyridoxal phosphate, the purification process resulted in decreased activity unless this cofactor was added (table 3). Moreover, the absolute requirement for the cofactor is indicated by the effects of pyridoxal inhibitors as shown in table 4. At 25 mM, hydroxylamine, thiosemicarbazide, INAH (isonicotinamide hydrazide), and D-cycloserine all caused significant inhibition.

The effect of compounds structurally related to ornithine on the activity of OTA was also determined. Significant inhibition was caused by norvaline, δ-aminovaleric acid, and α-ketovaleric acid. Inhibitory effects by hydroxyproline, isoleucine, leucine, proline, and valine were not detected (table 4).

Since the optimum concentration for α-ketoglutarate in the fluke OTA reaction was much higher than that of rat liver (Strecker, 1965) and rabbit liver (Ertel and Isseroff, unpublished), the α-ketoglutarate concentration in *Fasciola* was determined. In *Fasciola* the level of α-ketoglutarate was found to be about 214 nanomoles/g wet weight (table 5), or approximately 4.0 $\times$ greater than the concentration of that compound in rat liver.

TABLE 4

EFFECT OF VARIOUS SUBSTANCES ON THE
ACTIVITY OF ORNITHINE-δ-TRANSAMINASE

Ornithine, α-ketoglutarate, and pyridoxal-5-phosphate were present in all
incubations at 35.0, 20.0, and 2.5 μmoles/ml, respectively. The tested
substances were present at 25.0 μmoles/ml. The pH was 8.0 in all vessels.
Values listed are the means of five samples. Inhibitions listed were significant
at $P < 0.05$ in a two tailed t test.

Addition	μmoles PCA produced/mg of protein/hour ± S.E.	% Inhibition
none	22.90 ± 0.728	—
L-hydroxyproline	22.69 ± 0.915	0
L-proline	21.05 ± 0.538	0
L-arginine	24.05 ± 0.919	0
L-isoleucine	20.70 ± 0.994	0
L-leucine	21.72 ± 0.467	0
L-valine	20.64 ± 1.011	0
norvaline	15.78 ± 1.523	31
δ-aminovaleric acid	17.53 ± 1.048	23
α-ketovaleric acid	18.56 ± 1.261	19
hydroxylamine	none detected	100
D-cycloserine	13.35 ± 1.053	42
thiosemicarbazide	none detected	100
INAH (isonicotinamide hydrazide)	9.39 ± 1.188	59

TABLE 5

COMPARISON OF α-KETOGLUTARATE CONCENTRATION IN
FASCIOLA AND RAT LIVER

Values are in nanomoles of α-ketoglutarate/gram wet weight and each is
based on six samples.

Tissue	α-ketoglutarate concentration ± S.E.	Significant difference from Fasciola
Fasciola	214 ± 10.0	—
rat liver	54 ± 10.0	$P < 0.001$

We undertook additional studies to test the effectiveness of amino group acceptors other than α-ketoglutarate. In several publications, Kurelec (1974, 1975a, 1975b) has suggested that an ornithine-alanine transaminase is active in the production of proline by *Fasciola*. If so, pyruvate ought to function as an amino group acceptor. Using the purified enzyme and standard enzyme assay conditions (35 mM ornithine, 2.5 mM pyridoxal PO_4, pH 8), we found that neither pyruvate, oxaloacetate, nor glyoxylate (each at 20 mM), is effective in producing measurable PCA. We also examined the possibility that ornithine-alanine transaminase activity might be present in the crude homogenate. Numerous experiments, such as those shown in table 6, indicated that α-ketoglutarate is much more effective than the other amino group acceptors tested and that pyruvate activity can be demonstrated only at concentrations that probably are beyond the physiological range.

DISCUSSION

The pattern of inhibition of *Fasciola* OTA by various structurally related compounds was found to be similar to that reported by Strecker (1965) for rat liver, except that valine, leucine, and isoleucine caused no significant inhibition. Complete inhibition of *Fasciola* OTA was obtained with some inhibitors of pyridoxal-dependent enzymes, indicating that this enzyme, like other transaminases, requires this cofactor. Furthermore, the purification procedure, which undoubtedly resulted in a loss of endogenous pyridoxal, caused a decrease in enzyme activity, which could then be partially restored by the addition of pyridoxal-5-phosphate.

The K_m for ornithine in purified preparations of *Fasciola* OTA is 2.33 mM, a figure similar to that found for rat liver OTA (Strecker, 1965). However, the K_m of α-ketoglutarate for *Fasciola* (2.24 mM) is much higher than that of α-ketoglutarate in rat liver (0.28 mM) (Strecker, 1965). Fluke OTA, there-

TABLE 6

DATA FROM REPRESENTATIVE EXPERIMENTS COMPARING ACTIVITY OF
ORNITHINE-δ-TRANSAMINASE IN CRUDE HOMOGENATES OF *FASCIOLA*
WITH VARIOUS AMINO GROUP ACCEPTORS

Ornithine and pyridoxal-5-phosphate were present at 35.0 and 2.5 μmoles/ml, respectively. Specific activity values are in μmoles of PCA/mg protein/hour. Each value is the mean of five samples.

| Acceptor | Specific activity $\pm$ S.E. at various millimolar acceptor conc. | | |
	3.75 mM	10.0 mM	75.0 mM
α-ketoglutarate	2.74 ± 0.076	3.45 ± 0.082	not tested
glyoxylate	none detected	0.92 ± 0.043	9.89 ± 0.061
pyruvate	none detected	none detected	0.62 ± 0.070

fore, appears to have a much greater requirement for α-ketoglutarate than either rabbit or rat liver. While mammalian liver OTA may be inhibited at concentrations lower than 10 mM α-ketoglutarate, inhibition of the fluke OTA enzyme was not observed even at 50 mM. It is likely, therefore, that in the fluke the reaction proceeds in the presence of high concentrations of α-ketoglutarate. Measurements made in *Fasciola* support such a conclusion. The source of the α-ketoglutarate may be the highly active glutamic-pyruvic transaminase (GPT) that was demonstrated to occur in *Fasciola* by Connelly and Downey (1968). This enzyme in the worm has eighteen times the activity of its counterpart in liver. The high GPT activity may explain Kurelec's (1974, 1975a) observation that the addition of either arginine or ornithine to *Fasciola* homogenates resulted in increased levels of alanine. Kurelec, however, interpreted his data as indicating that the amino group acceptor for the ornithine transaminase in PCA formation was pyruvate. To support this postulation he presented additional data (Kurelec, 1975a) indicating that the order of formation of compounds involved in proline biosynthesis was proline, alanine, and glutamate. The pattern suggested to Kurelec that the path of the amino group of ornithine is from ornithine to alanine to glutamate. However, in the same paper he reported that α-ketoglutarate was more effective in stimulating proline synthesis than pyruvate when either compound was added with ornithine to *Fasciola* homogenates.

To the present investigators, Kurelec's observations are not inconsistent with α-ketoglutarate as the amino group acceptor in the reaction whereby PCA is produced from ornithine. Furthermore, our data, based on direct measurement of enzyme activity, show that pyruvate is a very poor substrate for PCA formation in *Fasciola*.

Kurelec's observations can be explained by assuming that a highly active GPT, such as that described above, is coupled to the OTA reaction. The GPT would regenerate α-ketoglutarate from the glutamate produced in the OTA reaction. If sufficient ornithine, pyruvate and α-ketoglutarate were available, large amounts of alanine would be produced because of the repeated reconversion of glutamate into α-ketoglutarate in order to meet the high keto acid requirements of the OTA reaction. If this situation occurred in *Fasciola*, the high levels of free alanine found in the worm by Kurelec and Rijavec (1966) would also be explained.

We conclude that *Fasciola* OTA appears to be similar to rat liver OTA in most of the properties studied, except that the fluke enzyme has a greater requirement for α-ketoglutarate. Thus in *Fasciola* it is not only the high specific activity of OTA, but also the high tolerance of this enzyme for α-ketoglutarate, which results in high levels of proline formation.

ACKNOWLEDGMENT

The research for this paper was supported in part by grant AI-09911 from NIH-NIAID.

108 RICE UNIVERSITY STUDIES

REFERENCES CITED

Bergmeyer, H. U. and E. Bernt
1965 2-Oxoglutarate. *In* Methods of Enzymatic Analysis, H. U. Bergmeyer, ed. New York: Academic Press. Pp. 324-327.

Connelly, J. F. and N. E. Downey
1968 Glutamate transaminase activities of the liver fluke, *Fasciola hepatica* (Linnaeus, 1758). Research in Veterinary Science **9**:248-250.

Ertel, J. and H. Isseroff
1974 Proline in Fascioliasis: I. Comparative activities of ornithine-δ-transaminase and proline oxidase in *Fasciola* and in mammalian livers. Journal of Parasitology **60**:574-577.

Isseroff, H. and C. P. Read
1969 Studies on membrane transport: VI. Absorption of amino acids by fascioliid trematodes. Comparative Biochemistry and Physiology **30**:1153-1159.

Isseroff, H., M. Tunis, and C. P. Read
1972 Changes in amino acids of bile in *Fasciola hepatica* infections. Comparative Biochemistry and Physiology **41B**:157-163.

Jenkins, W. T. and H. Tsai
1970 Ornithine aminotransferase (pig kidney). *In* Methods in Enzymology, H. Tabor and C. W. White, eds., Vol. XVII A. New York: Academic Press. Pp. 281-285.

Katunuma, N., Y. Matsuda, and I. Tomino
1964 Studies on ornithine-ketoacid transaminase: I. Purification and properties. Journal of Biochemistry **45**:500-503.

Kurelec, B.
1974 Die physiologische Funktion der Arginase im Leberegel (*Fasciola hepatica*, L.). Acta Parasitologica Iugoslavica **5**:33-43.

1975a Catabolic path of arginine and NAD regeneration in the parasite *Fasciola hepatica*. Comparative Biochemistry and Physiology **51B**:151-156.

1975b Molecular biology of helminth parasites. International Journal of Biochemistry **6**:375-386.

Kurelec, B. and M. Rijavec
1966 Amino acid pool of the liver fluke (*Fasciola hepatica* L.). Comparative Biochemistry and Physiology **19**:525-531.

Lineweaver, H. and D. Burk
    1934 Determination of enzyme dissociation constants. Journal of the
    American Chemical Society 56:658-666.

Lowry, O. H., N. J. Rosebrough, A. L. Farr, and R. J. Randall
    1951 Protein measurement with the Folin phenol reagent. Journal of
    Biological Chemistry 193:265-275.

Strecker, H. J.
    1965 Purification and properties of rat liver ornithine-δ-transaminase.
    Journal of Biological Chemistry 240:1225-1230.

ULTRASTRUCTURAL CYTOCHEMISTRY OF THE TEGUMENTAL SURFACE MEMBRANE OF *PARAGONIMUS KELLICOTTI*

by Francis M. Gress and Richard D. Lumsden

ABSTRACT

The body surface of adult *Paragonimus kellicotti* is a syncytial epithelium bounded on its free surface by a trilaminar plasmalemma. The outer layer of this membrane is invested with a hirsute coat which, on the basis of its cytochemical staining properties with concanavalin A, bismuth subnitrate, ruthenium red, polycationic ferritin, and colloidal iron, appears to be rich in acidic carbohydrate. Morphological association and common cytochemical reactivity between Golgi complexes of tegumentary cytons and membranous granular inclusions provide evidence for the incorporation of carbohydrate into the granule matrix within the Golgi apparatus. The observed fusion of such inclusions with the body surface suggests a mechanism whereby the contents of cytoplasmic granules contribute to the maintenance and/or renewal of the carbohydrate-rich surface coat and indicates that the glycocalyx of the *P. kellicotti* tegument is not simply a layer of adsorbed host mucin. Although the binding of extrinsic (i.e., host) glycan to the body surface is not excluded by these observations, it appears that the cytochemically demonstrable glycocalyx that invests this trematode consists, at least in part, of the carbohydrate moieties of membrane macromolecules. Ionized acidic moieties associated with the carbohydrate-rich surface coat contribute substantially to the body surface electronegative charge and may partially account for the ability of the flukes to survive in the immunocompetent host.

Francis Gress is N.I.H. Predoctoral Fellow at Tulane University. Richard Lumsden is Professor of Biology and Dean of the Graduate School at Tulane University.

Consideration of the molecular composition, physico-chemical properties, and physiological activities of cell surface membranes has provided investigators with new insight into the probable organization of these supramolecular complexes (Stoeckenius and Engelman, 1969; Dewey and Barr, 1970; Singer and Nicolson, 1972; Oseroff et al. 1973; Singer, 1974). A recent advance of major significance has been the realization that carbohydrate-containing molecules are important constituents of cell surface membranes (Cook, 1968). The results of cytochemical, biochemical, and immunological studies, when considered collectively, indicate that much of the carbohydrate associated with the cell surface is a constituent of the membrane proper and as such does not constitute a superficial and extraneous coat of adsorbed extrinsic polysaccharide (Cook, 1968; Martinez-Palomo, 1970; Winzler, 1970; Kraemer, 1971; Cook and Stoddart, 1973). Sugar moieties present are covalently bonded to membrane proteins and lipids (Cook, 1968; Winzler, 1972; Hughes, 1973) and constitute a histochemically demonstrable glycocalyx (Bennett, 1963; Rambourg, 1971).

Numerous functions have been attributed to the chemical constituents of the glycocalyx. Membrane-associated carbohydrates are known to constitute the functional determinants of a variety of cell surface antigens (Winzler, 1972; Hughes, 1973) and otherwise influence cell surface antigenicity (Apffel and Peters, 1970); ionized acidic sugar moieties contribute substantially to the negative surface charge present on cell types thus far examined (Cook, 1968; Cook and Stoddart, 1973; Weiss, 1973); and surface-associated carbohydrate is believed to exert important control over the social behavior of cells (Roseman, 1970; Cook and Stoddart, 1973; Emmelot, 1973; Hakomori, 1973).

Examination of the body surfaces of a variety of parasitic platyhelminths using cytochemical, autoradiographic, and biochemical techniques has provided results consistent with the presence of a glycocalyx in association with the tegumental surface membrane (reviewed by Lumsden, 1975). Physiological and immunological considerations would suggest that detailed information concerning the nature of the body surface membrane and associated carbohydrate coat would provide a better understanding of the nature of the host-parasite relationship.

A considerable amount of information is available detailing the pathological and serological status of the host in paragonimiasis (Sadun et al., 1959; Capron et al., 1965; Yogore et al., 1965; Yokogawa, 1965; Seed et al., 1966 and 1968; Tada, 1967; Lumsden and Sogandares-Bernal, 1970; Chung, 1971; Sogandares-Bernal and Seed, 1973), but little is known concerning the physiology of the trematode itself. In the case of pulmonary disease, adult worms may survive for several or many years. The infection assumes a chronic course and inflammatory changes transform airways harboring adult worms into fibrotic pulmonary lesions or cyst-like structures. Moreover, sera of infested hosts are known to contain antibodies reactive against worm antigens. Adult worms, however, at least those from primary infections, are not rejected by host defense mechanisms. We report here observations on the body surface fine structure and topochemistry of *Paragonimus kellicotti* that may relate to the failure to reject and to other features of host-parasite interaction in pulmonary paragonimiasis.

MATERIALS AND METHODS

Adult *Paragonimus kellicotti* were removed from the lungs of domestic cats three or four years following the administration of 15 to 30 metacercariae. The worms were fixed in 5.0% glutaraldehyde in Millonig's phosphate buffer containing 3.0% sucrose and washed in buffer containing 5.0% sucrose. For ultrastructural study, material was post-fixed in Millonig's phosphate buffered 1.0% osmium tetroxide containing 2.0% sucrose, dehydrated with ethanol, and embedded in epoxy resin (Lumsden, 1970). Sections were stained with aqueous 2.0% uranyl acetate and lead citrate (Reynolds, 1963) and examined in a Siemens 1A electron microscope operated at an accelerating voltage of 80kv.

Cytochemical procedures included incubation of glutaraldehyde-fixed material in ruthenium red (Luft, 1971) and polycationic ferritin (Miles-Yeda Ltd., Rehovoth, Israel) according to the method of Danon et al. (1972). Staining with cationic and anionic iron colloids was performed according to methods described by Gasic et al. (1968). Procedures intended to reduce surface negative charge prior to cytochemical staining included incubation of tissue in neuraminidase (Sigma Chemical Co., type V; 4 hour incubation at 37°C in 0.05M acetate buffer, pH5.5, containing 0.85% NaCl and 10.0 units/ml of neuraminidase), poly-L-lysine-HBr (New England Nuclear, molecular weight 160,000; 1 hour incubation in phosphate buffered saline, pH7.2, containing 10.0mg/ml of poly-lysine), and methanolic-HCl (Lillie, 1954). Additional material was incubated in native (i.e., anionic) ferritin (K&K Laboratories; 1 hour incubation in phosphate buffered saline, pH7.2, containing 10.0mg/ml of ferritin).

Methods employed for the detection of concanavalin A binding sites were those previously described by Stein and Lumsden (1973) and McCracken and Lumsden (1975). In brief, specimens previously exposed to this lectin were incubated in a solution of horse radish peroxidase, transferred to a mixture of diaminobenzidene and hydrogen peroxide, and then osmicated. Control procedures included incubation of concanavalin A-labeled tissue in 0.1M alpha-methyl-D-mannoside prior to transfer to peroxidase and completion of the reaction sequence. Other material was incubated in ferritin-conjugated concanavalin A in phosphate buffered saline (Nicolson and Singer, 1971).

Thin sections collected on nickel or steel grids were treated with periodic acid (0.8gm dissolved in 100ml of 70% ethanol containing 0.02M sodium acetate) followed by alkaline bismuth subnitrate (Ainsworth et al., 1972) in order to detect 1,2-glycols of polysaccharide containing macro-molecules. Control procedures included blocking of periodate-engendered aldehydes with 1.0M m-aminophenol prior to treatment with alkaline bismuth subnitrate, removal of osmium from tissue sections with 1.0% hydrogen peroxide, and omission of the oxidation step from the reaction sequence.

RESULTS

Electron microscopic examination revealed that the body surface of *Paragonimus kellicotti* adults is a cytoplasmic syncytium containing numerous mitochondria and discoidal, membrane-bound inclusions (figure 1). Cytoplasmic processes connect this protoplasmic surface with nucleated tegumentary cytons located within the deeper parenchymal tissues of the worm (figure 2). The perinuclear cytoplasm is typically filled with inclusions similar to those observed in the surface syncytium (figures 2 and 3). Also present are mitochondria, a well developed granular endoplasmic reticulum, Golgi membranes, and free ribosomes (figure 3). The folded body surface is bounded by a trilaminar plasmalemma, the outer dense aspect of which is invested with a thin hirsute coat (figure 4).

Incubation of tissue in ruthenium red resulted in the formation of a uniform and continuous layer of reaction product along the outer surface of the tegument. The ruthenium-osmium complex was localized at the outer aspect of the surface membrane and was confined to the region occupied by the hirsute coat seen on conventionally stained material (figure 5). Examination of unstained thin sections through the body surface revealed that certain of what appear to be vesicular cytoplasmic inclusion bodies reacted with ruthenium red as well (figure 6). Since ruthenium red does not typically cross intact membranes, however (Luft, 1971), these may represent cross sectioned canalicular extensions of the free surface plasmalemma.

Polycationic ferritin employed at pH 7.2 uniformly covered the body surface and was bound by the hirsute coat associated with the surface plasmalemma (figure 7). Methylation or treatment of tissue with poly-lysine prior to incubation in the positively charged ferritin resulted in a considerable reduction in the amount of polycationic ferritin subsequently bound (figure 8).

Uniform and dense staining of the body surface membrane was obtained when cationic colloidal iron was employed at pH 2.0 (figures 9 and 10). Prior incubation of material in poly-lysine largely prevented the subsequent binding of positively charged iron micelles (figures 11 and 12). It is to be noted as well that the integrity of the surface membrane was preserved following treatment with poly-lysine (figure 12). Essentially no binding of anionic colloidal iron was observed when this stain was employed at pH 5.5. Treat-

FIGURES 1-12 ON FOLLOWING PAGES.

FIG. 1. SURVEY MICROGRAPH OF ADULT *Paragonimus kellicotti* body wall illustrating fine structural features of the tegument. Note the numerous mitochondria and discoidal inclusion bodies. The free tegumental surface is folded, and cell boundaries are absent. Section stained with uranyl acetate and lead citrate. ×28,000.

FIG. 2. SURVEY MICROGRAPH OF BODY WALL illustrating several tegumentary cytons (TC). The cytoplasm of tegumental cell bodies is filled with discoidal granules (gr). Section stained with uranyl acetate and lead citrate. ×10,500.

FIG. 3. CYTOPLASMIC ORGANELLES OF TEGUMENTAL PERIKARYA including mitochondria (m), free ribosomes (r), granular endoplasmic reticulum (er), and Golgi membranes (go). Discoidal granules (gr) are also present. (Nucleus, N). Section stained with uranyl acetate and lead citrate. ×36,000.

FIG. 4. TRILAMINAR BODY SURFACE PLASMA MEMBRANE. Note the hirsute coat associated with the outer dense leaflet of the membrane. Section stained with uranyl acetate and lead citrate. ×210,000.

FIG. 5. TEGUMENTAL SURFACE STAINED WITH RUTHENIUM RED. Note that opaque reaction product is localized at the outer surface of the plasmalemma and occupies the region of the hirsute coat. Section stained with uranyl acetate and lead citrate. ×129,000.

FIG. 6. TISSUE INCUBATED *EN BLOC* IN RUTHENIUM RED. Cytoplasmic inclusion bodies (*) different from the discoidal granules (gr) are stained by this cytochemical reagent. Section otherwise unstained. ×57,000.

FIG. 7. BINDING OF POLYCATIONIC FERRITIN TO THE OUTER FACE of the body surface membrane is uniform and dense. Incubation at pH 7.2. Section stained with uranyl acetate and lead citrate. ×76,500.

FIG. 8. SPECIMEN TREATED WITH METHANOLIC HCl prior to incubation in polycationic ferritin. Note the decrease in binding of cationic ferritin when compared with results illustrated in figure 7. Section stained with uranyl acetate and lead citrate. ×115,000.

FIG. 9. SURVEY MICROGRAPH ILLUSTRATING UNIFORM AND DENSE STAINING following treatment with cationic colloidal iron at pH 2.0. Section stained with uranyl acetate and lead citrate. ×33,800.

FIG. 10. TISSUE STAINED WITH CATIONIC COLLOIDAL IRON AT pH 2.0. Note that the iron micelles are bound to the outer dense leaflet of the plasmalemma. Section stained with uranyl acetate and lead citrate. ×185,000.

FIG. 11. SPECIMEN TREATED WITH POLY-L-LYSINE prior to staining with positively charged colloidal iron. Note the considerable reduction in binding of iron micelles when compared with results illustrated in figures 9 and 10. Section stained with uranyl acetate and lead citrate. ×57,500.

FIG. 12. MATERIAL TREATED WITH POLY-L-LYSINE prior to staining with cationic colloidal iron. The surface coat appears thickened, presumably due to the presence of electrostatically bound poly-lysine. Morphological integrity of the surface plasmalemma is preserved. Section stained with uranyl acetate and lead citrate. ×167,000.

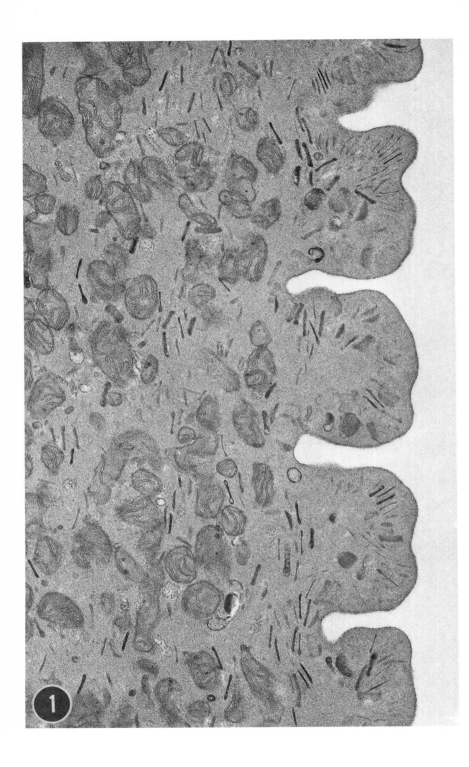

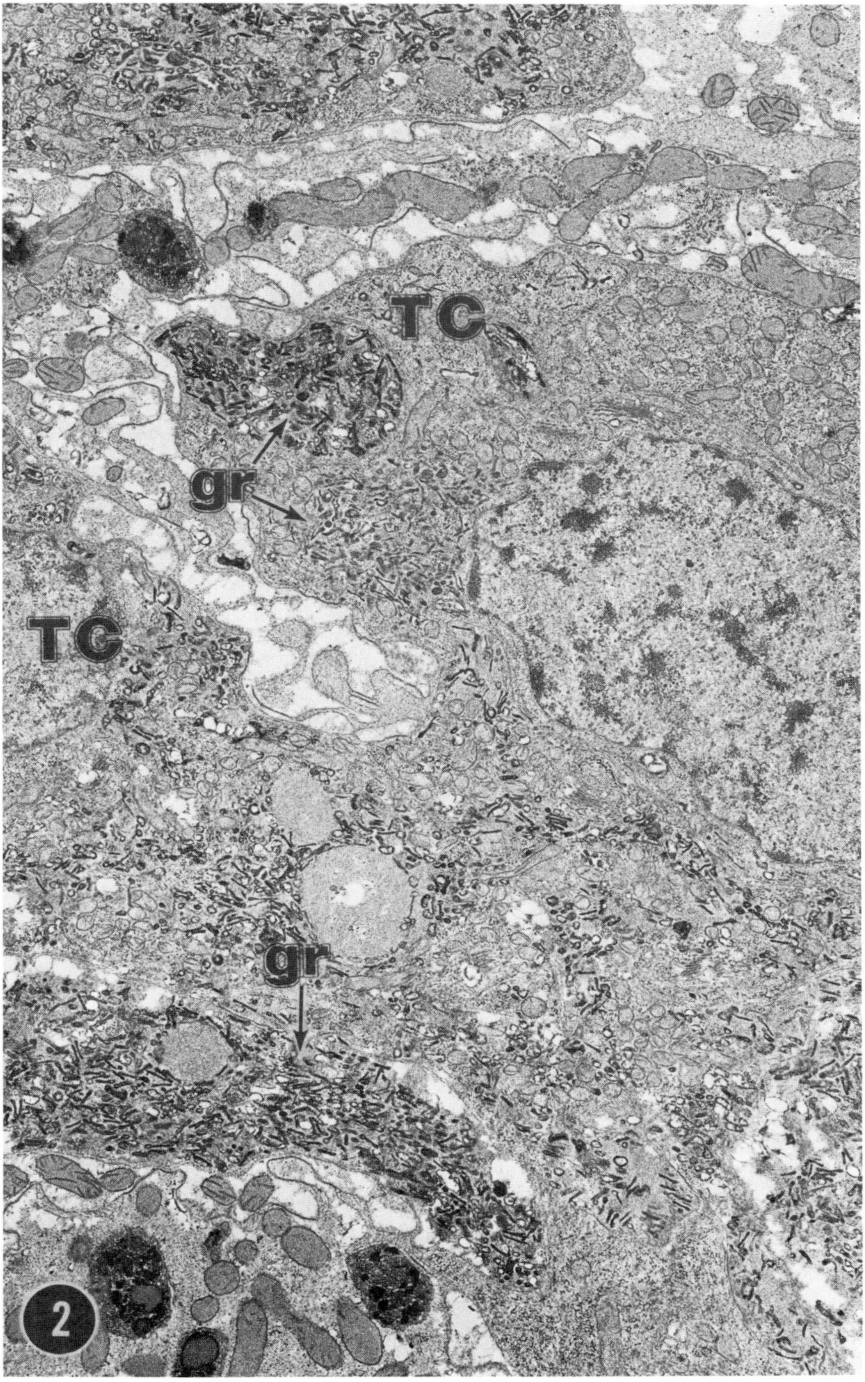
TC
gr
TC
gr
2

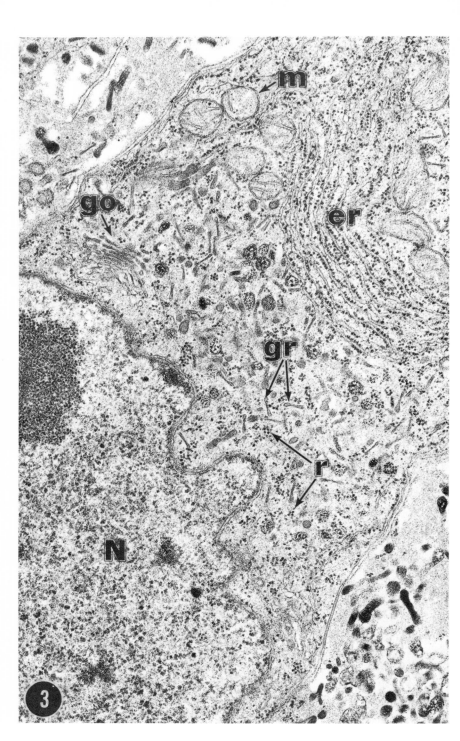

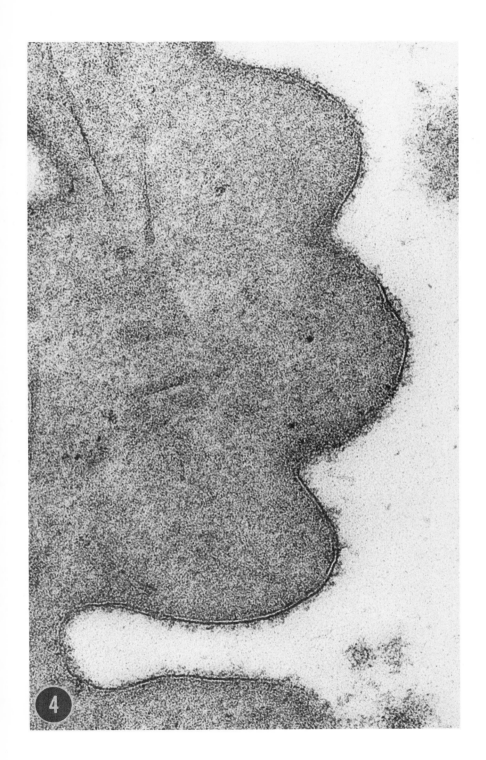

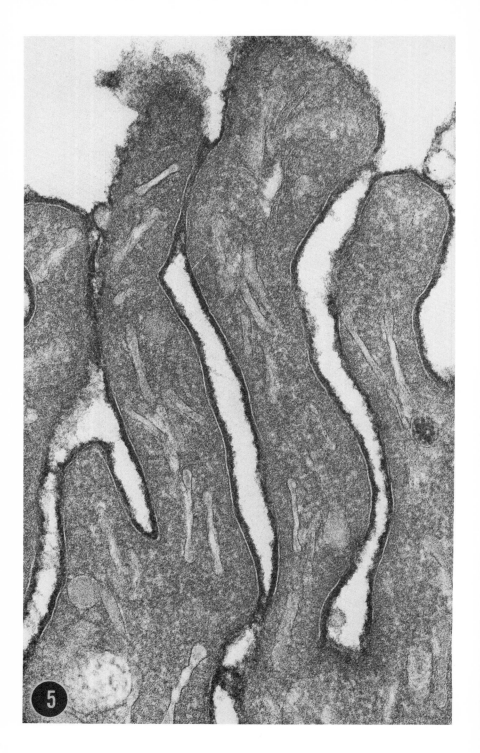

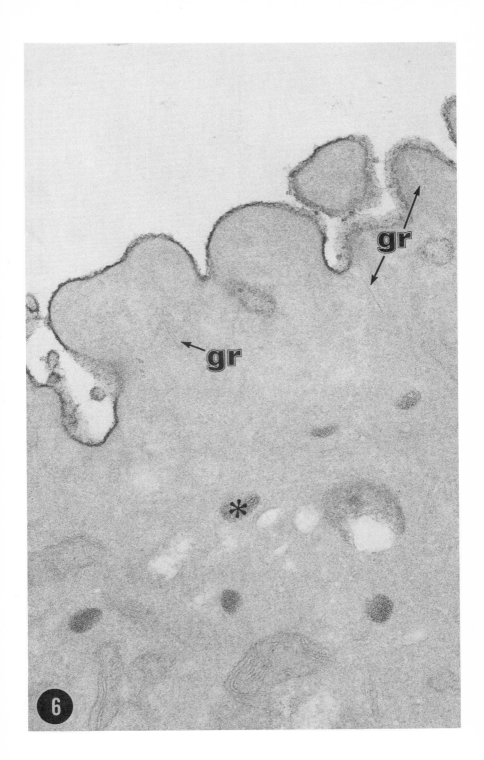

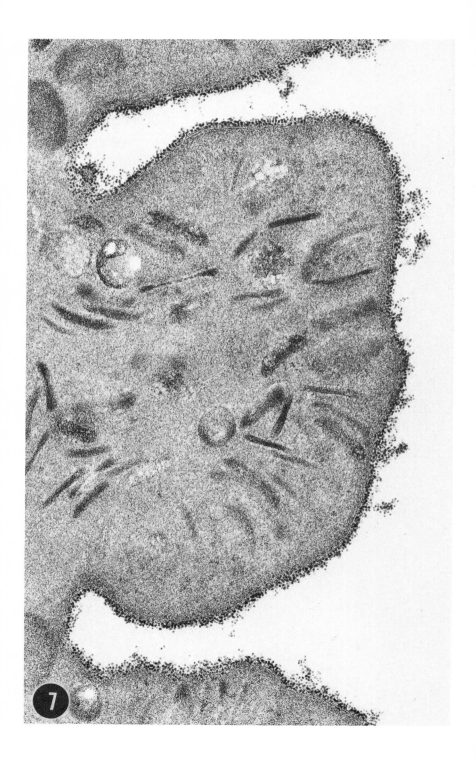

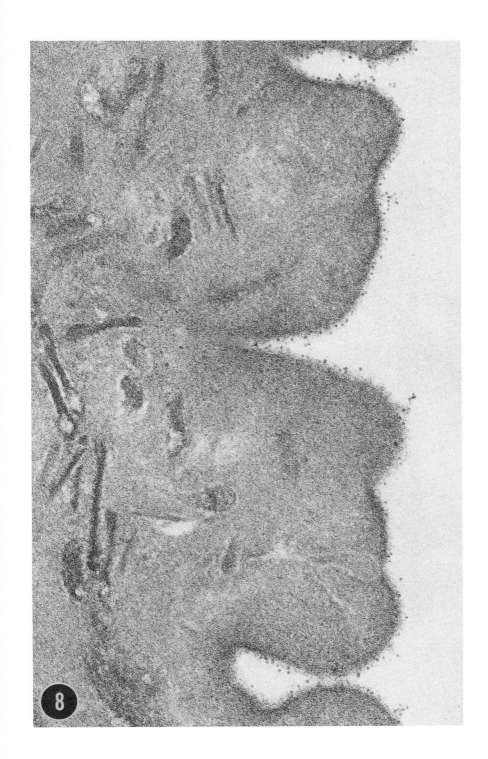

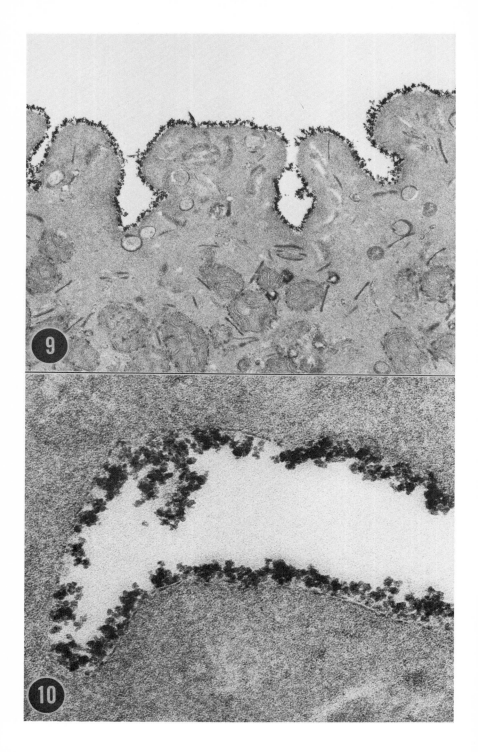

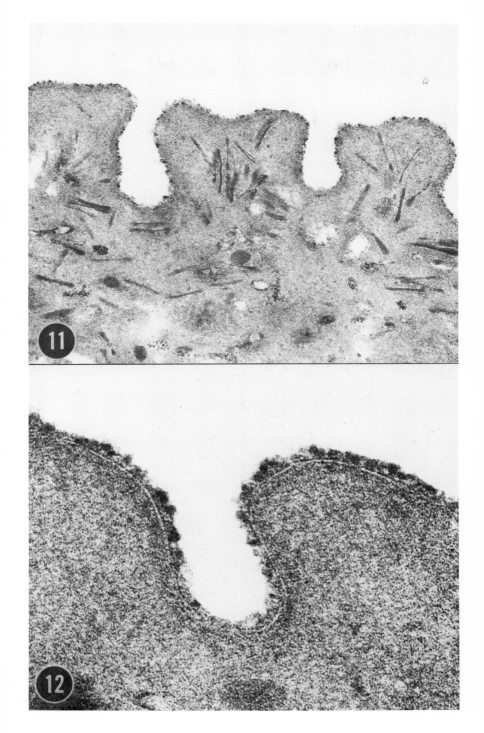

ment of tissue with neuraminidase did not unequivocally reduce the staining by cationic iron nor did it substantially increase the affinity of the surface membrane for the anionic colloid.

Binding of native (i.e., anionic) ferritin to the body surface was negligible and was not increased following methylation.

The osmium black reaction product of the concanavalin A-horse radish peroxidase-diaminobenzidene procedure was visualized as a continuous and dense precipitate associated with the outer leaflet of the plasmalemma (figures 13 and 14). The specificity of the procedure was confirmed by the observation that no precipitate was visible when concanavalin A-treated tissue was incubated in methyl-mannoside prior to completion of the reaction sequence with peroxidase, diaminobenzidene, and osmium tetroxide. Lectin binding sites were directly made visible by treating samples with ferritin-conjugated concanavalin A (figure 15).

The tegumental surface membrane and cytoplasmic inclusions were uniformly stained by reduced bismuth subnitrate following periodate oxidation (figure 16). Golgi membranes and inclusions present within perikarya reacted similarly (figures 17, 18, 19, and 20). Images such as those illustrated in figures 17, 19, and 20 suggest a secretory mechanism whereby budding vesicles limited by membranes derived from Golgi saccules undergo a condensation process resulting in the formation of the discoidal granules present in the perikarya and tegumental syncytium.

Survey micrographs illustrate palisade-like arrays of these membrane bound inclusions adjacent to the tegumental surface plasmalemma (figure 1), and examination of tissue sections that had been oxidized and treated with bismuth reagent revealed a continuity of cytochemically reactive substance(s) between some of these inclusions and the tegumental surface (figure 21).

Blockage of periodate-engendered aldehydes with m-aminophenol prevented the reduction of bismuth subnitrate. In addition, specific bismuth staining did not occur when the periodate oxidation step was omitted. Nonspecific bismuth staining was prevented when sections were treated with 1.0% hydrogen peroxide in order to remove osmium from the tissue.

DISCUSSION

The body surface of adult *Paragonimus kellicotti* is a syncytial epithelium consisting of an outer protoplasmic layer continuous with nucleated cytons located in the underlying parenchyma. This cyto-architecture conforms to that previously described for the tegument of other digenetic trematodes (reviewed by Lee, 1966 and 1972; Inatomi et al., 1970; Lumsden, 1975).

The free tegumental surface of *P. kellicotti* is bounded by a trilaminar plasmalemma. The outer layer of this membrane is invested with a thin hirsute coat, which appears by its cytochemical staining properties to be rich in acidic glycans. The presence of carbohydrate in this coat is specifically

FIGURES 13-21 ON FOLLOWING PAGES.

FIG. 13. ELECTRON-DENSE REACTION PRODUCT uniformly coats the body surface following treatment of concanavalin A-incubated tissue with peroxidase and diaminobenzidene. Section stained with uranyl acetate and lead citrate. ×72,600.

FIG. 14. OSMIUM BLACK REACTION PRODUCT localizes lectin binding sites to the outer surface of the tegumental surface membrane following treatment of concanavalin A-incubated tissue with peroxidase and diaminobenzidene. Section stained with uranyl acetate and lead citrate. ×336,000.

FIG. 15. DIRECT LOCALIZATION OF CONCANAVALIN A BINDING SITES is achieved by incubating tissue in ferritin-conjugated lectin. Note that lectin binding sites are associated with the body surface hirsute coat. Section stained with uranyl acetate and lead citrate. ×177,000.

FIG. 16. SURVEY MICROGRAPH ILLUSTRATING UNIFORM STAINING of the tegumental surface and cytoplasmic inclusion bodies by bismuth subnitrate following periodate oxidation. Section otherwise unstained. ×65,000.

FIG. 17. FOLLOWING PERIODATE OXIDATION, Golgi saccules (go), budding vesicles (v), and cytoplasmic granular inclusions (gr) of the tegumental perikarya are stained by reduced bismuth subnitrate. Section otherwise unstained. ×92,400.

FIG. 18. CYTOPLASMIC GRANULES OF TEGUMENTAL PERIKARYA stain specifically following treatment with periodic acid and bismuth subnitrate. Section otherwise unstained. ×98,500.

FIG. 19. IMAGES SUCH AS THE ONE ILLUSTRATED HERE suggest a secretory mechanism whereby vesicles (v) derived from Golgi saccules (go) condense to form recognizable discoidal granules (gr). Note the variations in opacity of the condensing bodies and the close association of granular inclusions with the Golgi apparatus. Bismuth reagent stains the matrices of granular inclusions as well as membranes of Golgi saccules and vesicles. Section treated with periodate and alkaline bismuth subnitrate. ×71,300.

FIG. 20. GOLGI APPARATUS OF TEGUMENTAL PERIKARYON illustrating specific bismuth staining of Golgi membranes, budding and condensing vesicles, and granular inclusions. Continuity of a vesicular body (v) with a granular inclusion suggests that vesicles derived from Golgi saccules condense to form the carbohydrate-containing granular inclusions. Section treated with periodate followed by bismuth subnitrate. ×61,800.

FIG. 21. REDUCED BISMUTH SUBNITRATE is deposited over the tegumental surface (*) following periodate oxidation. Cytoplasmic discoidal granules (gr) stain specifically with bismuth reagent as well. The continuity of cytochemically demonstrable material between the tegumental surface and granular inclusion illustrated here suggests that these carbohydrate-containing granules may contribute their contents to the glycocalyx which invests the body surface of *P. kellicotti*. Section otherwise unstained. ×184,800.

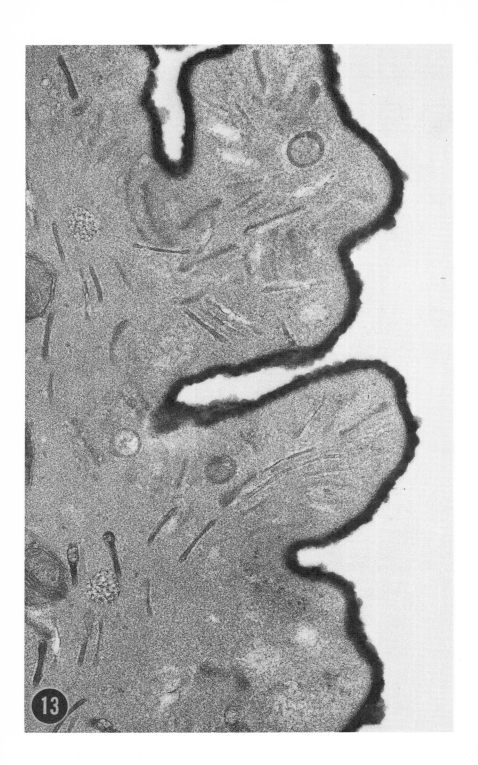

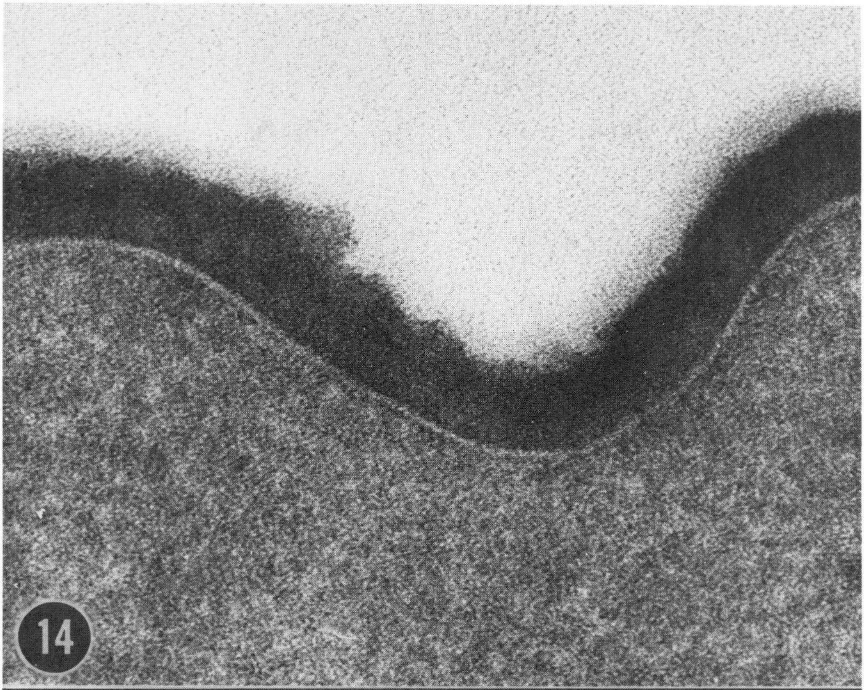
14

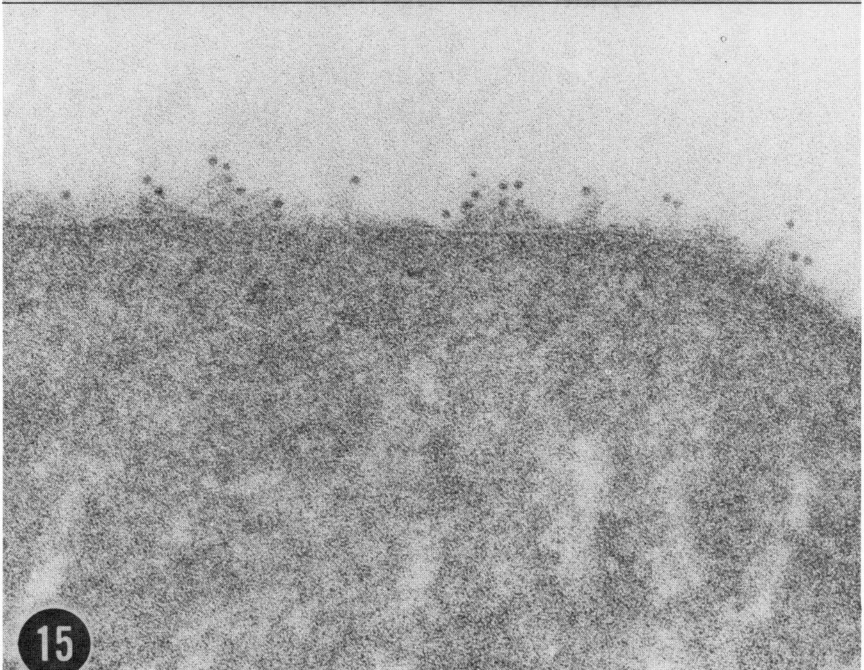
15

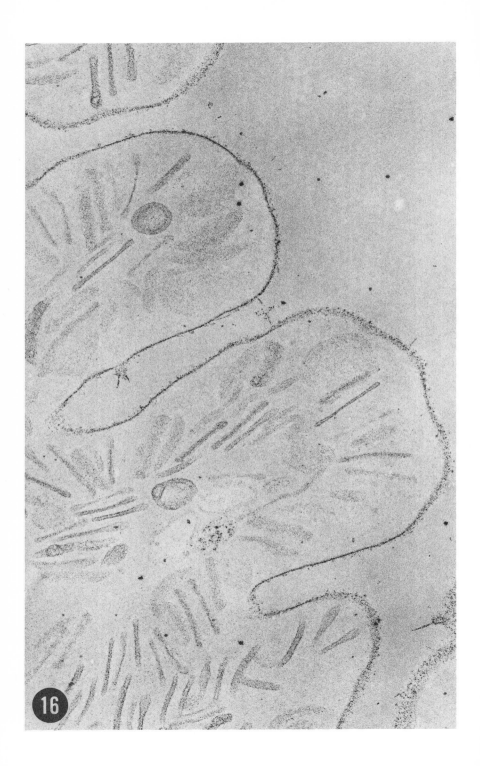

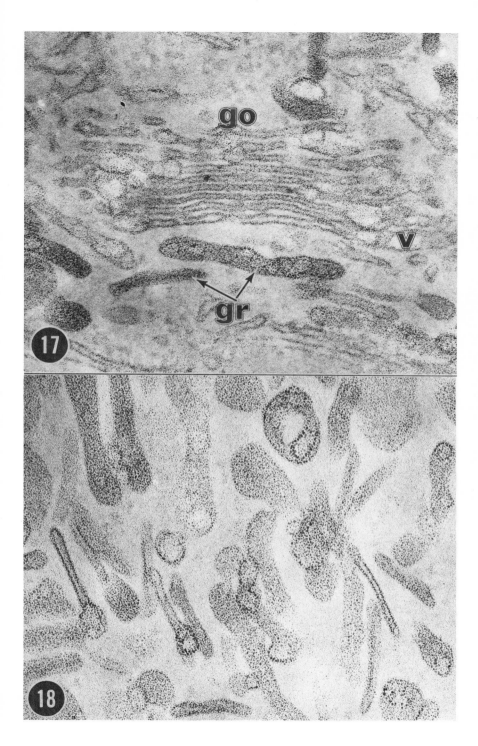

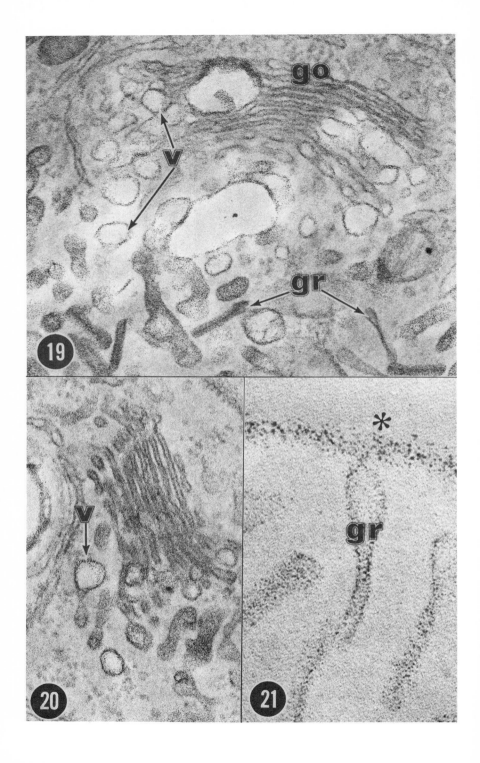

indicated by its affinity for the lectin concanavalin A, which binds to sugars having the D-arabinopyranoside configuration at C_3, C_4, and C_6 (Lis and Sharon, 1973). As indicated by the use of ferritin or peroxidase-conjugated concanavalin A, lectin binding sites at the tegumental surface of *P. kellicotti* are numerous, uniformly distributed, and associated with the hirsute coat. Further confirmation of the carbohydrate content of this coat was obtained with the periodic acid-bismuth subnitrate technique of Ainsworth et al. (1972), which is considered to yield results for electron microscopy comparable to those obtained with the periodate-Schiff procedure employed for light microscopic demonstration of carbohydrate-containing macromolecules.

The polyanionic nature of the surface coat is indicated by its staining with cationic ferritin, colloidal iron, and ruthenium red. Pretreatment of tissue with poly-lysine greatly reduced staining by cationic iron and ferritin, presumably because this strongly basic polypeptide electrostatically neutralized many of the acidic substances associated with the body surface (Mamelak et al., 1969; Lumsden et al., 1970; Lumsden, 1972). Furthermore, treatment of tissue with methanolic HCl, which destroys sulfate groups and esterifies carboxylic acid functions (Lillie, 1954), caused a marked reduction in the binding of colloidal metal cations, presumably due to a net reduction in surface electronegativity. Attempts to stain the surface with anionic colloidal iron and native ferritin (negatively charged at neutral pH) proved unsuccessful. The failure to stain might be explained by the occurrence of a charge repulsion and/or masking effect due to the presence of coat-associated anions (Lumsden et al., 1970) or to a paucity of cationic constituents accessible to the reagents employed. Indicative of the latter possibility is the observation that methylation procedures did not convincingly increase the binding of negative colloidal iron or anionic ferritin.

Considered collectively, the results of cytochemical tests employing ruthenium red, polycationic ferritin, and cationic colloidal iron indicate that acidic carbohydrate is a constituent of the body surface membrane hirsute coat. At the low pH used for staining with cationic colloidal iron, the ionization of free carboxyl groups of proteins is suppressed, whereas more strongly acidic sulfate, carboxyl, and phosphate groups bound to carbohydrate or lipid may remain ionized (see Lumsden et al., 1970; Lumsden, 1973, for discussion) and may thus account for the body surface electronegativity. Also, we regard as untenable the view that the cell surface negative charge may be attributed in its entirety to membrane phospholipids (Seaman and Heard, 1960; Wallach and Eylar, 1961; Eylar et al., 1962; Cook, 1968; McLaughlin et al., 1971).

Numerous studies (see, for example, Cook and Stoddart, 1973; Weiss, 1973) have demonstrated that neuraminic (sialic) acids are important ionogenic species at the surface of animal cells. Sialic acid is a constituent of membrane glycoprotein and glycolipid (Cook, 1968), and the carboxyl group of this acidic sugar (pK_a 2.6) is responsible for a significant portion of the negative surface charge of animal cells (Wallach and Eylar, 1961; Eylar et al., 1962; Cook, 1968). Our attempts to detect the presence of sialic acid at the tegumental surface of *P. kellicotti* (by demonstration of a reduction in staining with cationic iron and polycationic ferritin following treatment of material with neuraminidase) were inconclusive. Furthermore, no substantial increase in the binding of anionic iron was observed following enzymatic digestion. Such qualitative studies do not, however, rule out the possibility that sialic acid may be present, since the incubation medium was not assayed for enzymatically liberated sugar. Lumsden et al. (1970) noted that chemical fixation interfered with subsequent enzymatic removal of carbohydrate from the body surface of the cestode *Hymenolepis diminuta*, and it is conceivable that cross-linking of membrane constituents by the glutaraldehyde used in the present study might have induced conformational changes, thereby rendering the alpha ketoside linkage of sialic acid non-labile to the action of neuraminidase. It has been noted as well that certain sialic acid residues bound to glycoprotein resist cleavage by neuraminidase (Labat and Schmid, 1969). Furthermore, sialic acid bound to glycolipid is not susceptible to hydrolysis by neuraminidase (Weinstein et al., 1970). The possibility that a substantial portion of the surface negative charge may be due to the presence of membrane-bound ionogenic species other than the carboxyl groups of neuraminic acid, such as sulfate and phosphate, must also be considered (Hauser et al., 1969; Monis et al., 1969; Rothman and Elder, 1970; Allen, Winzler et al., 1971; Allen, Ault, et al., 1974).

Membranous granular inclusions of the type seen in the tegument of *P. kellicotti* are ubiquitous among trematodes and cestodes, and have been implicated in several diverse functions (see Lumsden, 1975, for review). The morphological associations between the Golgi apparatus and these inclusions and their common staining with bismuth subnitrate following periodate oxidation suggest that carbohydrate, probably as glycoprotein, is incorporated into the granule matrix via the Golgi apparatus of the tegumentary cytons. The presence of palisaded accumulations of these granules immediately below the tegument surface membrane and the apparent fusion of such granules with it would suggest a mechanism whereby the contents of such granules may contribute to the maintenance or renewal of the glycocalyx. Molecular constituents of a wide variety of cell surface membranes, including those of other parasitic helminths, undergo continuous turnover (Warren and Glick, 1968; Oaks and Lumsden, 1971; Hughes et al., 1972), the carbohydrate components in many cases being elaborated in the form of membrane-limited vesicles originating from the Golgi apparatus (see, for example, Bennett and Leblond, 1970; Bennett et al., 1974).

The failure of the tegument granules of *P. kellicotti* to stain for the acidic moieties so demonstrated in the glycocalyx is not inconsistent with this view.

It is to be noted that staining with ruthenium red, colloidal iron, and ferritin was carried out *en bloc* rather than applied to sections, and these substances typically do not cross intact membranes. Nor were the granules stained in the tissues incubated *en bloc* with concanavalin A. This lack of staining is also most likely due to the inability of the concanavalin A and/or the ferritin/peroxidase markers to pass through the tegument plasmalemma and vesicle membranes, rather than to an absence of lectin binding material in the vesicles.

It would appear that the glycocalyx of the *P. kellicotti* tegument is not simply a layer of adsorbed host mucins. Repeated washing failed to reduce subsequent staining with the cytochemical reagents employed, and there is evidence (noted above) for the elaboration of surface glycans by the worms. However, binding of extrinsic (i.e., host) glycans is not excluded by these observations. A glycan sharing antigenic specificity with the AB antigens of host erythrocytes has been identified in the surface coat of schistosomula and adults of *Schistosoma mansoni* (Clegg, 1972). These as well as Forssman antigens (which are glycolipids) have been shown by Dean and Sell (1972) and by Dean (1974) to be passively adsorbed by these worms *in vitro*.

The similarities in topochemistry between adult *P. kellicotti* and *S. mansoni* are noteworthy in considering the functional implications of the *Paragonimus* glycocalyx. As noted by Stein and Lumsden (1973) and Lumsden (1975), the highly acidic glycocalyx of the blood-dwelling schistosomes would be expected to militate against their entrapment by clot formation and inflammatory elements. The pulmonary capsules in which the adults of *P. kellicotti* reside are highly vascularized and the central space contains an exudate of plasma and extravasated formed elements. The flukes' survival in this environment may be due, at least in part, to the electronegative potential of the body surface, which could serve to repel similarly charged inflammatory or immunocompetent cells (Currie and Bagshawe, 1967; Hause et al., 1970), platelets, and fibrinogen (Sawyer and Pate, 1953; Sawyer, Pate, and Weldon, 1953; Mattson and Smith, 1973). The possibility that acidic carbohydrates may reduce the immunogenicity of body surface constituents or otherwise impair the functioning of humoral factors produced by the immune system of the host (Apffel and Peters, 1970) must also be considered. Whether or not potentially protective host antigenic determinants (Clegg, 1972) or blocking antibodies (Feldman, 1972) may coat the body surface as extrinsic components of the glycocalyx remains to be established.

Little information is presently available regarding the biochemistry of *Paragonimus* (see Hamajima, 1971, 1972, 1973a, and 1973b), the nutritional requirements of these flukes, and the possibility that the tegumental membrane may function as an absorptive surface. Evidence that the activities of hydrolytic enzymes and transport systems associated with the body surfaces of enteric parasitic platyhelminths are dependent upon cationic cofactors (see, for example, Dike and Read, 1971; Lumsden, 1973; Lumsden and

Berger, 1974) would suggest that hydrophilic and acidic carbohydrate moieties present at the body surface might function to bind essential ions or water molecules and so facilitate surface enzyme or transport activity. Similar functional roles have been attributed to carbohydrates present at the surfaces of other cell types (Seaman et al., 1969; Langer and Frank, 1972).

ACKNOWLEDGMENT

Research was supported by grants (AI 08673 and AI 0002) from the National Institutes of Health.

REFERENCES CITED

Ainsworth, S. K., S. Ito, and M. J. Karnovsky
 1972 Alkaline bismuth reagent for high resolution ultrastructural demonstration of periodate-reactive sites. Journal of Histochemistry and Cytochemistry 20:995-1005.

Allen, H. J., C. Ault, R. J. Winzler, and J. F. Danielli
 1974 Chemical characterization of the isolated cell surface of *Amoeba*. Journal of Cell Biology 60:26-38.

Allen, H. J., R. J. Winzler, C. Ault, and J. F. Danielli
 1971 Studies on the anionic nature of cell surface of *Amoeba discoides*. Abstracts of the 11th Annual Meeting of the American Society for Cell Biology 9.

Apffel, C. A. and J. H. Peters
 1970 Regulation of antigenic expression. Journal of Theoretical Biology 26:47-59.

Bennett, G. and C. P. Leblond
 1970 Formation of cell coat material for the whole surface of columnar cells in the rat small intestine, as visualized by radioautography with L-fucose-^{3}H. Journal of Cell Biology 46:409-416.

Bennett, G., C. P. Leblond, and A. Haddad
 1974 Migration of glycoprotein from the Golgi apparatus to the surface of various cell types as shown by radioautography after labeled fucose injection into rats. Journal of Cell Biology 60:258-284.

Bennett, H. S.
 1963 Morphological aspects of extracellular polysaccharides. Journal of Histochemistry and Cytochemistry 11:14-23.

Capron, A., M. Yokogawa, J. Biguet, M. Tsuji, and G. Luffau
 1965 Diagnostic immunologique de la paragonimose humaine. Mise en

evidence d'anticorps sériques spécifiques par immunoelectro-phorèse. Bulletin de la Société de Pathologie Exotique **58**:474-487.

Chung, C. H.
1971 Human paragonimiasis. *In* Pathology of Protozoal and Helminthic Diseases. R. A. Marcial-Rojas, ed. Baltimore: Williams and Wilkins Company.

Clegg, J. A.
1972 The schistosome surface in relation to parasitism. Symposium of the British Society for Parasitology **10**:23-40.

Cook, G. M. W.
1968 Glycoproteins in membranes. Biological Reviews of the Cambridge Philosophical Society **43**:363-391.

Cook, G. M. W. and R. W. Stoddart
1973 Surface Carbohydrates of the Eukaryotic Cell. New York: Academic Press.

Currie, G. A. and K. D. Bagshawe
1967 The masking of antigens on trophoblast and cancer cells. The Lancet **1**:708-710.

Danon, D., L. Goldstein, Y. Marikovsky, and E. Skutelsky
1972 Use of cationized ferritin as a label of negative charges on cell surfaces. Journal of Ultrastructure Research **38**:500-510.

Dean, D. A.
1974 *Schistosoma mansoni*: adsorption of human blood group A and B antigens by schistosomula. Journal of Parasitology **60**:260-263.

Dean, D. A. and K. W. Sell
1972 Surface antigens on *Schistosoma mansoni*. II. Adsorption of a Forssman-like host antigen by schistosomula. Clinical and Experimental Immunology **12**:525-540.

Dewey, M. M. and L. Barr
1970 Some considerations about the structure of cellular membranes. Current Topics in Membranes and Transport **1**:1-33.

Dike, Sue Carlisle and C. P. Read
1971 Tegumentary phosphohydrolases of *Hymenolepis diminuta*. Journal of Parasitology **57**:81-87.

Emmelot, P.
1973 Biochemical properties of normal and neoplastic cell surfaces; a review. European Journal of Cancer **9**:319-333.

Eylar, E. H., M. A. Madoff, O. V. Brody, and J. L. Oncley
	1962 The contribution of sialic acid to the surface charge of the erythro-
	cyte. Journal of Biological Chemistry **237**:1992-2000.

Feldman, J. D.
	1972 Immunological enhancement: A study of blocking antibodies.
	Advances in Immunology **15**:167-214.

Gasic, G. J., L. Berwick, and M. Sorrentino
	1968 Positive and negative colloidal iron as cell surface electron stains.
	Laboratory Investigation **18**:63-71.

Hakomori, S.
	1973 Glycolipids of tumor cell membranes. Advances in Cancer Re-
	search **18**:265-315.

Hamajima, F.
	1971 Studies on metabolism of lung flukes, genus *Paragonimus*. IV.
	Reactions of glycolysis in homogenates of eggs, larvae, and adults.
	Japanese Journal of Parasitology **20**:475-480.

	1972 Studies on metabolism of lung flukes, genus *Paragonimus*. V.
	Reactions of the tricarboxylic acid cycle in homogenates of eggs,
	larvae, and adults. Japanese Journal of Parasitology **21**:280-285.

	1973a Studies on metabolism of lung flukes of the genus *Paragonimus*.
	VI. Succinoxidase system in homogenates of eggs, larvae, and
	adults. Experimental Parasitology **33**:515-521.

	1973b Studies on metabolism of lung fluke genus *Paragonimus*. VII.
	Action of bithionol on glycolytic and oxidative metabolism of
	adult worms. Experimental Parasitology **34**:1-11.

Hause, L. L., R. A. Pattillo, A. Sances, and R. F. Mattingly
	1970 Cell surface coatings and membrane potentials of malignant and
	nonmalignant cells. Science **169**:601-603.

Hauser, H., D. Chapman, and R. M. C. Dawson
	1969 Physical studies of phospholipids. XI. Ca^{2+} binding to monolayers
	of phosphatidylserine and phosphatidylinositol. Biochimica et
	Biophysica Acta **183**:320-333.

Hughes, R. C.
	1973 Glycoproteins as components of cellular membranes. Progress in
	Biophysics and Molecular Biology **26**:189-268.

Hughes, R. C., B. Sanford, and R. W. Jeanloz
	1972 Regeneration of the surface glycoproteins of a transplantable

mouse tumor cell after treatment with neuraminidase. Proceedings of the National Academy of Sciences U.S.A. **69**:942-945.

Inatomi, S., D. Sakumoto, Y. Tongu, S. Suguri, and K. Itano
1970 Electron Micrographs of Helminths. The 20th Anniversary Publication. Department of Parasitology, Okayama Medical School, Okayama, Japan.

Kraemer, P. M.
1971 Complex carbohydrates of animal cells: biochemistry and physiology of the cell periphery. Biomembranes **1**:67-190.

Labat, J. and K. Schmid
1969 Neuraminidase-resistant sialyl residues of α_1-acid glycoprotein. Experientia **25**:701.

Langer, G. A. and J. S. Frank
1972 Lanthanum in heart cell culture. Effect on calcium exchange correlated with its localization. Journal of Cell Biology **54**:441-455.

Lee, D. L.
1966 The structure and composition of the helminth cuticle. Advances in Parasitology **4**:187-254.

1972 The structure of the helminth cuticle. Advances in Parasitology **10**:347-379.

Lillie, R. D.
1954 Histopathologic Technic and Practical Histochemistry. New York: McGraw Hill.

Lis, H. and N. Sharon
1973 The biochemistry of plant lectins (Phytohemagglutinins). Annual Review of Biochemistry **42**:541-574.

Luft, J. H.
1971 Ruthenium red and violet. II. Fine structural localization in animal tissues. Anatomical Record **171**:369-415.

Lumsden, R. D.
1970 Preparatory technique for electron microscopy. *In* Experiments and Techniques in Parasitology. A. MacInnis and M. Voge, eds. San Francisco: Freeman Press.

1972 Cytological studies on the absorptive surfaces of cestodes. VI. Cytochemical evaluation of electrostatic charge. Journal of Parasitology **58**:229-234.

1973 Cytological studies on the absorptive surfaces of cestodes. VII.

Evidence for the function of the tegument glycocalyx in cation binding by *Hymenolepis diminuta*. Journal of Parasitology **59**:1021-1030.

1975 Surface ultrastructure and cytochemistry of parasitic helminths. Experimental Parasitology **37**:267-339.

Lumsden, R. D. and B. Berger
1974 Cytological studies on the absorptive surfaces of cestodes. VIII. Phosphohydrolase activity and cation adsorption in the tegument brush border of *Hymenolepis diminuta*. Journal of Parasitology **60**:744-751.

Lumsden, R. D., J. A. Oaks, and W. L. Alworth
1970 Cytological studies on the absorptive surfaces of cestodes. IV. Localization and cytochemical properties of membrane-fixed cation binding sites. Journal of Parasitology **56**:736-747.

Lumsden, R. D. and F. Sogandares-Bernal
1970 Ultrastructural manifestations of pulmonary paragonimiasis. Journal of Parasitology **56**:1095-1109.

Mamelak, M., S. L. Wissig, R. Bogoroch, and I. S. Edelman
1969 Physiological and morphological effects of poly-*L*-lysine on the toad bladder. Journal of Membrane Biology **1**:144-176.

Martinez-Palomo, A.
1970 The surface coats of animal cells. International Review of Cytology **29**:29-75.

Mattson, J. S. and C. A. Smith
1973 Enhanced protein adsorption at the solid-solution interface: dependence on surface charge. Science **181**:1055-1057.

McCracken, R. O. and R. D. Lumsden
1975 Structure and function of parasite surface membranes. II. Concanavalin A adsorption by the cestode *Hymenolepis diminuta* and its effect on transport. Comparative Biochemistry and Physiology **52B**:331-337.

McLaughlin, S. G. A., G. Szabo, and G. Eisenman
1971 Divalent ions and the surface potential of charged phospholipid membranes. Journal of General Physiology **58**:667-687.

Monis, B., A. Candiotti, and J. E. Fabro
1969 On the glycocalyx, the external coat of the plasma membrane, of some secretory cells. Zeitschrift für Zellforschung **99**:64-73.

Nicolson, G. L. and S. J. Singer
  1971 Ferritin-conjugated plant agglutinins as specific saccharide stains for electron microscopy. Application to saccharides bound to cell membranes. Proceedings of the National Academy of Sciences U.S.A. **68**:942-945.

Oaks, J. A. and R. D. Lumsden
  1971 Cytological studies on the absorptive surfaces of cestodes. V. Incorporation of carbohydrate-containing macromolecules into tegument membranes. Journal of Parasitology **57**:1256-1268.

Oseroff, A. R., P. W. Robbins, and M. M. Burger
  1973 The cell surface membrane; biochemical aspects and biophysical probes. Annual Review of Biochemistry **42**:647-682.

Rambourg, A.
  1971 Morphological and histochemical aspects of glycoproteins at the surface of animal cells. International Review of Cytology **31**:57-114.

Reynolds, E. S.
  1963 The use of lead citrate at high pH as an electron opaque stain in electron microscopy. Journal of Cell Biology **17**:208-212.

Roseman, S.
  1970 The synthesis of complex carbohydrates by multiglycosyl-transferase systems and their potential function in intercellular adhesion. Chemistry and Physics of Lipids **5**:270-297.

Rothman, A. H. and J. E. Elder
  1970 Histochemical nature of an acanthocephalan, a cestode, and a trematode absorbing surface. Comparative Biochemistry and Physiology **33**:745-762.

Sadun, E. H., A. A. Buck, and B. C. Walton
  1959 The diagnosis of paragonimiasis westermani using purified antigens in intradermal and complement fixation tests. Military Medicine **124**:187-195.

Sawyer, P. N. and J. W. Pate
  1953 Bio-electric phenomena as an etiologic factor in intravascular thrombosis. American Journal of Physiology **175**:103-107.

Sawyer, P. N., J. W. Pate, and C. S. Weldon
  1953 Relations of abnormal and injury electric potential differences to intravascular thrombosis. American Journal of Physiology **175**:108-112.

Seaman, G. V. F. and D. H. Heard
1960 The surface of the washed human erythrocyte as a polyanion. Journal of General Physiology 44:251-268.

Seaman, G. V. F., P. S. Vassar, and M. J. Kendall
1969 Calcium ion binding to blood cell surfaces. Experientia 25:1259-1260.

Seed, J. R., F. Sogandares-Bernal, and A. A. Gam
1968 Studies on American paragonimiasis. VI. Antibody response in three domestic cats infected with *Paragonimus kellicotti*. Tulane Studies in Zoology and Botany 15:70-74.

Seed, J. R., F. Sogandares-Bernal, and R. R. Mills
1966 Studies on American paragonimiasis. II. Serological observations of infected cats. Journal of Parasitology 52:358-362.

Singer, S. J.
1974 The molecular organization of membranes. Annual Review of Biochemistry 43:805-833.

Singer, S. J., and G. L. Nicolson
1972 The fluid mosaic model of the structure of cell membranes. Science 175:720-731.

Sogandares-Bernal, F. and J. R. Seed
1973 American paragonimiasis. Current Topics in Comparative Pathobiology 2:1-56.

Stein, P. C. and R. D. Lumsden
1973 *Schistosoma mansoni*: topochemical features of cercariae, schistosomula, and adults. Experimental Parasitology 33:499-514.

Stoeckenius, W. and D. M. Engelman
1969 Current models for the structure of biological membranes. Journal of Cell Biology 42:613-646.

Tada, I.
1967 Physiological and serological studies of *Paragonimus miyazakii* infection in rats. Journal of Parasitology 53:292-297.

Wallach, D. F. H. and E. H. Eylar
1961 Sialic acid in the cellular membranes of Ehrlich ascites-carcinoma cells. Biochimica et Biophysica Acta 52:594-596.

Warren, L. and M. C. Glick
1968 Membranes of animal cells. II. The metabolism and turnover of the surface membrane. Journal of Cell Biology 37:729-746.

Weinstein, D. B., J. B. Marsh, M. C. Glick, and L. Warren
  1970 Membranes of animal cells. VI. The glycolipids of the L cell and its surface membrane. Journal of Biological Chemistry **245**:3928-3937.

Weiss, L.
  1973 Neuraminidase, sialic acids, and cell interactions. Journal of the National Cancer Institute **50**:3-19.

Winzler, R. J.
  1970 Carbohydrates in cell surfaces. International Review of Cytology **29**:77-125.

  1972 Glycoproteins of plasma membranes. Chemistry and function. *In* Glycoproteins. Their Composition, Structure, and Function. A. Gottschalk, ed. New York: Elsevier Publishing Company.

Yogore, M. G., R. M. Lewert, and E. D. Madraso
  1965 Immunodiffusion studies on paragonimiasis. American Journal of Tropical Medicine and Hygiene **14**:586-591.

Yokogawa, M.
  1965 *Paragonimus* and paragonimiasis. Advances in Parasitology **3**:99-158.

THE EFFECT OF CORTISONE ON THE SURVIVAL OF *HYMENOLEPIS DIMINUTA* IN MICE

by C. A. Hopkins and Helen E. Stallard

ABSTRACT

Cortisone acetate, 1 mg thrice weekly, permitted *Hymenolepis diminuta* in single worm infections to survive without loss for 45 days in male CFLP mice 6 weeks old at infection. Worms grew to over 200 mg dry weight by day 18 post infection (PI), following which their weight remained steady or decreased slightly. In control mice, worms rarely exceeded 12 mg before being rejected on day 11 ± 2 PI. Termination of cortisone (commenced on day 1 of the infection) on day 6, 8, 10, or 20 PI led in each case to loss of worms starting six days later, which is believed to be 2-3 days after the cortisone given in the last dose had ceased to be effective. Tolerance was not induced. Delaying the commencement of cortisone until day 8 PI did not accelerate rejection after cortisone was terminated. Decreasing the cortisone, from 1 mg to 0.5 mg thrice weekly, after 20 days led to a reduction in the size of the worms, possibly because the worms experienced a host rejection attack when the cortisone level fell, between injections, below its protective level. There was abundant evidence that cortisone has an immediate protective effect on tapeworms, since even when cortisone treatment was delayed until after destrobilation, worms recovered and regrew as they would on transfer to a naive host. The possible sites of action of cortisone in preventing rejection are discussed.

INTRODUCTION

Six-week-old male CFLP mice reject *Hymenolepis diminuta* between day 8 and day 16 PI. The time decreases as the worm burden increases from a single to a 12-worm infection. The process usually involves worms destrobilating, leaving a scolex and neck 0.5-10 mm long, which are subsequently lost without regrowth (Hopkins, Subramanian, and Stallard, 1972a; Befus and

---

C. A. Hopkins is Professor of Zoology at the University of Glasgow. Helen Stallard is a Research Assistant at the University of Glasgow.

Featherston, 1974). Various immunosuppressants including cortisone prevent this rejection (Hopkins, Subramanian, and Stallard, 1972b). The purpose of the present work was to determine what stage(s) in the process of worm rejection was (were) blocked by cortisone—in particular, to determine how early in an infection it was necessary to start cortisone to prevent rejection, and whether the subsequent course of an infection was the same after cortisone treatment was terminated, following different periods of administration.

MATERIALS AND METHODS

CFLP male mice (from Anglia Laboratories, formerly Carworth Europe Ltd.) 39-45 days old at time of infection were used. A single cysticercoid of *H. diminuta* was administered by stomach tube to each mouse. Mice were kept 5 or 6 per cage. Except where specified to the contrary, 1 mg of cortisone acetate (Cortisyl Roussel) in 0.05 ml of 0.9% NaCl was administered intramuscularly into right and left hind limbs alternately on Mondays, Wednesdays, and Fridays starting on day 1 (mice infected Tuesday, day 0). Mice were killed and the small intestine was removed and examined for worms. In the absence of parasites, the intestine was cut into 10 cm sections, which were placed separately in 5 cm petri dishes containing Hanks' saline and incubated for $^1/_2$-2 hours at 37° C. Examination quickly revealed any worm bigger than 0.2 mg dry weight, i.e., over approximately 2 cm long, but more prolonged searching was necessary to find destrobilated worms, usually less than one cm in length. Destrobilated worms were not looked for, nor included when found by chance.

RESULTS

The effect of commencing cortisone at different stages of an infection

Cortisone treatment was commenced on day 1, 3, 6, 8, 10, 13, and 15 PI, in seven groups of mice to which a single cysticercoid per mouse had been administered on day 0 (Tuesday). Intramuscular injection of 1 mg of cortisone was given to each mouse on Monday, Wednesday, and Friday of each week until animals were autopsied. Mice were autopsied in groups of ten; the number infected and the mean dry weight of the worms, excluding worms < 0.2 mg, are shown in table 1. Figure 1 shows the biomass, i.e., total dry weight of worm recovered from ten mice, at different stages in the course of the infection; the various symbols indicate the day cortisone was started.

From table 1 and figure 1, it is apparent that it makes little difference whether cortisone is started 1, 3, or 6 days after infection. In each group (A, B, and C, table 1) over 90% of cysticercoids administered were recovered as worms, and in each group worm growth was at a decreasing exponential

TABLE 1

THE EFFECT OF COMMENCING CORTISONE AT DIFFERENT TIMES DURING AN INFECTION ON THE SURVIVAL AND GROWTH OF *H.DIMINUTA* IN MICE (SINGLE WORM INFECTIONS)

1 mg cortisone administered thrice weekly starting:

Day post infection	Group A Day 1 No./10	Mean wt.	Group B Day 3 No./10	Mean wt.	Group C Day 6 No./10	Mean wt.	Group D Day 8 No./10	Mean wt.	Group E Day 10 No./10	Mean wt.	Group F Day 13 No./10	Mean wt.	Group G Day 15 No./10	Mean wt.	Group H None No./10	Mean wt.
8	8	4	8	4	10	2	10	—	—		—		—		—	
9	10	11	9	8	10	7	7	4	—		—		—		6	2
10	10	24	10	18	8	8	8	12	—		—		—		5	8
12	10	63	10	46	10	34	8	39	—		—		—		3	9
14	9	106	8	88	9	110	10	71	3 + 0	75 + 0	1 + 0	86 + 0	—		1	21
16	10	229	10	225	10	168	9	174	2 + 3	109 + 0.4	1 + 0	95 + 0	1	0.2	1	0.9
18	10	255	8	262	10	234	9	194	4 + 1	168 + 3	1 + 2	246 + 3	1	2	2	0.7
20	9	234	9	244	9	235	10	179	3 + 4	209 + 6	2 + 2	170 + 1	1	3	0	
22	9	240	9	211	10	232	9	183	4 + 3	202 + 6	0 + 1	0 + 11	1	214	0	
24	10	212	10	194	10	222	10	165	6? + 2?	170 + 56	1 + 5	220 + 27	1	9	0	
Mean No. worms (days 16-24)	9.6		9.2		9.8		9.4		6.4		3.6		1.0		0.6	

No./10, number of mice infected in a group of 10, each mouse given 1 cysticercoid on day 0; mean wt, mean dry weight (mg) of the worms recovered; —, indicates no mice examined.

The worms in Groups E and F are separated into two categories (see text); the first figure refers to worms that did not destrobilate, the second figure to destrobilated worms that were regrowing under cortisone. The two categories could not be distinguished, with certainty, on day 24 in Group E. Bottom line shows the mean number of mice infected (number of worms found) out of 10 at autopsy on days 16-24 inclusive.

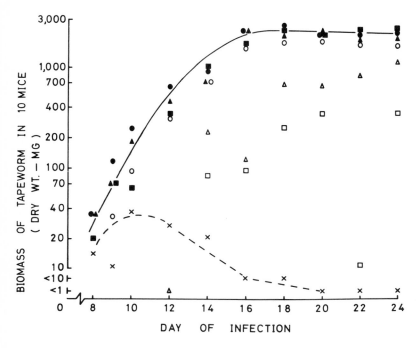

FIG. 1. THE CHANGE IN TAPEWORM BIOMASS during growth, in relation to the time when cortisone was commenced. Each point is the total dry weight of *H. diminuta* recovered from the
autopsy of ten mice given one cysticercoid each on day 0. Symbols indicate time when cortisone
was commenced: ● day 1, ▲ day 3, ■ day 6, o day 8, △ day 10, □ day 13, × Controls – no
cortisone. Curves fitted by eye, solid line based on closed points, broken line on crosses, i.e.,
worms in control mice. See text for heterogeneity of △ and □ data, and table 1 for further details.

rate until about day 18, after which a common plateau was maintained
(figure 1).

In Group D cortisone treatment was not commenced until day 8 PI, by
which time growth appears to have been retarded—cf. groups A and B. No
worm loss occurred, however (we recovered only 7 worms on day 9, almost
certainly because one or more small worms were missed; as growth proceeded
under cortisone the recovery rate rose to over 90%). The plot in figure 1
suggests that these worms, like those in Groups A, B, and C, reached a maximum size on day 18 and thereafter remained fairly constant. The implication
is that growth lost in the pre-patent period (eggs appear in the fèces on day
16 ± 1) is not made up.

In Group E (and F) the worm recoveries and mean weights are divided into
two categories (table 1). The first figure in each column refers to large worms
like those recovered in Groups A–D, the second figure to destrobilated
worms. We conclude:

(1) that approximately 25% of worms were rejected before cortisone had its effect (as shown by a 73% recovery on days 20-24 PI compared with over 90% in the Groups A-D)

(2) that approximately half the remaining worms destrobilated, but these were saved from expulsion and regrew. As they became bigger, the probability of finding them increased, and hence the total number of worms over 0.2 mg recovered increased from 3 on day 14 to 8 on day 24 PI.

Only 10% of the worms in Group F mice had not destrobilated and/or been expelled by the time the cortisone, started on day 13 PI, became effective. These worms, shown in the first column (table 1), persisted and reached a weight around 200 mg. As in Group E, destrobilated worms a few mm long and therefore not included started to regrow once cortisone was started. Five days after the commencement of cortisone administration, these destrobilated worms exceeded 0.2 mg (second column, Group F, table 1).

By day 15 PI, as can be seen in the controls (Group H), virtually all worms either had been rejected or were present only as destrobilated scolices. Even at this stage, however, the administration of cortisone can save destrobilated worms and permit them to re-initiate growth (Group G). The worm recovered on day 22 PI was clearly a worm in a slow or non-responder mouse (slow-responding mice, like this one in the Group G autopsied on day 22 PI, are occasionally found in control mice in similar experiments, although none happened to be found in the control mice in this experiment).

The rather low recovery of worms in the control mice (Group H) on day 9 PI was almost certainly due to the small size of the worms; rejection or destrobilation appears to have occurred on day 11 ± 2 PI.

Effect of terminating cortisone administration on the longevity of the tapeworm
One cyst of *H. diminuta* was administered to each of 430 mice on day 0. On day 1 the mice were divided into five groups: Group A, 100 control mice not receiving cortisone; Group B, 85 mice receiving cortisone on days 1, 3, and 6 PI; Group C, 75 mice receiving cortisone on days 1, 3, 6, and 8 PI; Group D, 65 mice receiving cortisone on days 1, 3, 6, 8 and 10 PI; Group E, 105 mice receiving cortisone throughout experiment, i.e., days 1, 3, 6, 8, 10, 13, 15, 17, 20, and 22 PI.

The number of worms (exceeding 0.2 mg) and their mean dry weight is shown in table 2. Each figure is based on a group of ten mice except those indicated with an asterisk, in which the number was reduced to nine because of mortality during the experiment. The results indicate that:

(1) In the control mice (Group A), worms were lost on day 11 ± 2 PI as expected.

(2) In Group E, the cortisone regimen prevented the rejection of worms throughout the 24 days of the experiment.

TABLE 2

The Effect of Terminating Cortisone, after Different Periods, on the Growth and Survival of Single Worm Infections of *H. diminuta* in SPF CFLP Male Mice

Age of infection (days)	Group A No Cortisone		Group B Cortisone to Day 6		Group C Cortisone to Day 8		Group D Cortisone to Day 10		Group E Cortisone Throughout	
	no. of mice infected/10	mean dry wt. of worm (mg)	no. of mice infected/10	mean dry wt. of worm (mg)	no. of mice infected/10	mean dry wt. of worm (mg)	no. of mice infected/10	mean dry wt. of worm (mg)	no. of mice infected/10	mean dry wt. of worm (mg)
6	—	—	—	—	—	—	—	—	—	—
8	10	1.7	—	—	—	—	—	—	9	2.8
9	8	2.2	—	—	—	—	—	—	9	7.0
10	5	2.7	8	3.6		—	—	—	6	9.3
12	1	9.5	5	12	9	26		—	9	42
14	1	0.9	3	27	7	53	9	43	7	65
16	0		2	24	0		5	115	8	152
18	0		0		1	123	5	102	7	203
20	0		0*		1	119	2*	131	8*	197
22	0		0*		2	77	0*		8*	202
24	0		1*	54	0		0*		7*	174

Group A mice received no cortisone, mice in Groups B, C, D, and E received cortisone starting on day 1 of the infection (see text) until and including the day specified. The horizontal line shows when the cortisone treatment was terminated. —, no mice examined; 0, none of the mice infected; *, 9 mice (not 10) in the group.

(3) In Groups B, C, and D, cessation of cortisone was followed by the same sequence of events: no loss of worms occurred in the subsequent 4 days, by day 6 PI worm rejection commenced, by day 8 50% or less of the worms remained, and by day 10 80% of the worms had been rejected.

(4) Cortisone treatment led not only to the survival of the worms but to continuous increase in worm weight until day 18 PI, at which stage weight increase stopped, presumably because worms had reached sexual maturity.

(5) The mean weight of control worms was less than that of worms of the same age from mice receiving cortisone (cf. Groups A and E). This result could occur either because cortisone increases the "normal rate" of growth (cf. growth of *H. microstoma* in mice—Moss, 1972) or because the immune response of the host had begun to affect worm growth by day 8 PI in the control mice.

Tolerance
Attempts were made to induce tolerance to worm infection under cortisone, which Wakelin and Selby (1974) have shown to occur in *Trichuris muris*.

Mice were divided into three groups. Group 1, infected day 0, received cortisone from day 1 through day 20 PI (●, figure 2). On day 21, cages were randomly divided into three subgroups a, b, and c. Mice in subgroup 1a continued to receive cortisone until the experiment was terminated on day 45 PI (●, day 23-45); in subgroup 1b the mice received no further treatment (o, figure 2); and in subgroup 1c the mice had the cortisone level reduced by half, i.e., to doses of 0.5 mg (■, figure 2). Group 2 mice were controls and received no cortisone (×, figure 2). Group 3 mice were on cortisone day 8-20 PI (▲, figure 2), after day 20 the results are shown as △, figure 2.

  The results show that:
(1) Following termination of cortisone on day 20 PI, rejection took place 6 ± 1 days later; it made no difference whether cortisone administration had been started on day 1 (o, figure 2) or day 8 (△, figure 2).

(2) One mg cortisone thrice weekly maintained worms without loss for 45 days (●, figure 2). Maximum weight, with maturation, was reached in the period day 16-21 PI and maintained until day 23; thereafter the worms were lighter. Without replication of the experiment, it is not possible to say whether or not the fall in mean worm weight from day 23 to day 28 PI followed by a rise is significant (cf. oscillation in weight of *Diphyllobothrium latum* in dogs [Wardle and Green, 1941], and growth of *H. citelli* [Hopkins and Stallard, 1974, figures 3 and 4]).

(3) When the level of cortisone was reduced to 0.5 mg/dose after day

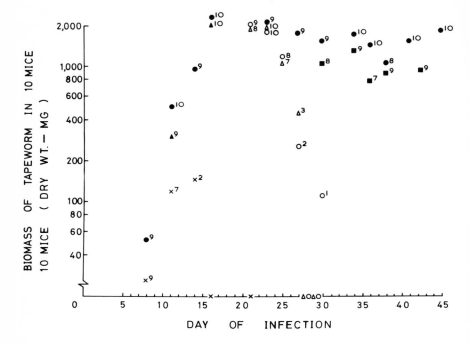

FIG. 2. THE EFFECT ON *H. diminuta* of terminating or decreasing the amount of cortisone on day 20 of an infection. Closed points indicate mice receiving cortisone up to autopsy; open points, mice in which cortisone was stopped on day 20; ×, control mice (no cortisone). Indices show number of mice in group of 10 infected (single worm infections). ●, full dose cortisone started day 1; ▲, full dose cortisone started day 8; ■, mice receiving half normal dose after day 20.

20 PI, the weight of the worms fell (■, figure 2) compared with those in mice kept on 1 mg/dose, and stabilized around 100 mg dry weight, compared with 150-190 mg in the mice on 1 mg/dose.

(4) In control mice worm growth was stunted by day 11 PI, and worm rejection was nearly complete by day 14 PI (×, figure 2).

DISCUSSION

Although many authors (e.g., Weinmann, 1966, 1970) have argued that parasitic worms living completely in the lumen of the intestine evoke immune responses, there is still a widely held belief that, since many intestinal tapeworms do not damage the gut mucosa, they are isolated outside the body of the host and are not immunogenic. Any lingering conjectures that intimate contact with the tissues (either during a migratory phase or following erosion of the mucosa by hooks on the scolex) is necessary before a protective immune response can be evoked, are no longer tenable (Hopkins, Subramian, and Stallard, 1972a and b; Wassom et al., 1974; Befus and Featherston, 1974; Hopkins and Stallard, 1974; Befus, 1975a). It is now well established that immunologically effective amounts of proteins are readily taken up through the intestinal wall of adult mammals (Bienenstock, 1974). The Thomas and Parrott (1974) have found that continued oral administration of small quantities of bovine serum albumin to mammals leads to partial immunological tolerance to this protein, an observation which is of particular relevance in intestinal parasitology, where potentially immunogenic helminths often avoid rejection by the host.

There is also no doubt that corticosteroids prevent the loss of worms from the intestine in numerous parasite infections (Wakelin, 1970). What is controversial is the point or points at which cortisone blocks the rejection mechanism. The problem is partly due to the multifaceted effects of cortisone (Claman, 1972, 1975; Berenbaum, 1974).

When cortisone was administered to mice commencing on day 1, 3, or 6 after infection, the survival and growth of the worms was similar (figure 1, table 1); presumably, therefore, the worms were up to this time unaffected by a host response. Delaying cortisone administration until day 8 PI showed that by this day worm growth had been affected; however, cortisone prevented further damage and permitted growth to resume (table 1). The most interesting result was that cortisone started on day 10, 13, or 15 PI—that is, after rejection had commenced—not only protected surviving intact worms but also permitted destrobilated worms to regrow, as they would if surgically transplanted into a previously uninfected host (Hopkins, Subramian, and Stallard, 1972a). These results resemble those obtained by Luffau and Urquhart (unpublished, but quoted by Jarrett et al., 1968) who found that cortisone started on day 14 PI stopped rejection of *Nippostrongylus brasiliensis* from rats.

It is known that *N. brasiliensis* is damaged by day 10 PI ("the antibody phase," Jones and Ogilvie, 1971) and that subsequent rejection occurs rapidly even after transferral into naive hosts, providing there are "normal" lymphocytes present (Keller and Keist, 1972). Cortisone is, therefore, when administered late in an infection, acting on this second phase of the rejection, or on a third component (Kelly et al., 1973) possibly activated by the T-lymphocytes involved in the second phase.

The evidence at present is too fragmentary to justify building a detailed hypothesis, but there are several similarities other than the production of antibody between the rejection of *H. diminuta* and *N. brasiliensis*. *H. diminuta* is coated with immunoglobulins by day 9 PI (Befus, 1974, 1975b), though this by itself is not likely to be the only factor in the rejection of *H. diminuta*, just as antibody is not the only factor in the rejection of *Nippostrongylus*. Indeed, antiworm antibodies are produced by rats infected with *H. diminuta* (Coleman et al., 1968; Harris and Turton, 1973), and mice with

H. microstoma (Moss, 1971; Goodall, 1973), in neither of which are worms rejected. Rejection of *H. diminuta*, like *Nippostrongylus*, is also dependent on the presence of T-cells as shown by Bland's (1976) work with "nude" mice. It is reasonable to assume that T-cells sensitized to *H. diminuta* exist long before day 10 of an infection, and as sensitized T-cells are cortisone resistant (reviewed, Claman, 1972) it seems most likely that cortisone is stopping rejection by blocking the action initiated by the sensitized T-cells.

As cortisone works extremely quickly in arresting rejection, it may be surmised that its effect is either on the surface of the tapeworm where the immunological attack probably occurs, or on a process occurring in the intestinal lumen or wall. What process(es) might be involved is mere speculation at present, for there are many possibilities. It is known, for instance, that cortisone affects the permeability of the host gut wall (Jarrett et al., 1968) and the release of prostaglandins (Lewis and Piper, 1975; Dineen et al., 1974), host responses which have been incriminated in the rejection of *Nippostrongylus*. It is reasonable to speculate that the ultimate factors that damage and/or cause the rejection of tapeworms, are enzymes. These enzymes could originate from complement, which Befus's (1975b) results using fluorescein-conjugated $\beta_1 C$ (C_3) antisera suggest is present on the surface of *H. diminuta* (cortisone is known to be anti-complementary—see Gewurz et al., 1965), or could be released from neutrophil polymorphs, which enter the intestine in large numbers when a sensitized mucosa is exposed to homologous antigen (Bellamy and Nielsen, 1974). The role of the T-cells may well be to stimulate this immune-mediated emigration, and hence the protective function of cortisone could be the stabilization of the lysosomal membranes, thus inhibiting release of the great variety of lytic enzymes that exist in polymorphs (Baggiolini, 1972).

Termination of cortisone, commenced on day 1 PI, on day 6, 8, or 10 of an infection gave similar results: rejection started 5-6 days later. However, it is known that 1.25 mg of cortisone acetate administered every three days prevents worm loss, although the weights of the worms are statistically significantly less than those of worms in mice given cortisone every two days; when it is administered every four days there is a significant loss of worms (Subramanian, unpublished). These results suggest that the effect of cortisone acetate lasts for about 72 hours after i.m. administration. Loss of worms, therefore, starts quickly, within two or three days of the effects of cortisone waning. This suggests that part or all of the induction phase of sensitization takes place under cortisone, and that it is the efferent or effector arms of the response that are blocked. Experiments to verify this by carrying out the whole primary infection under cortisone, terminating with "Zanil" (Hopkins, Grant, and Stallard, 1973), and then testing for memory by giving a secondary infection, have given inconclusive results. Two experiments showed a considerable reduction in the anamnestic response, whereas in a third experiment worms in the challenge infection were severely stunted, as in a normal secondary infection (Befus, 1975a).

One other result must fit into the rejection pattern. When cortisone was reduced on day 20 PI to half the normal dose, the weight of the worms in the mice decreased compared with the worms in mice kept on the full dose (figure 2). The process of rejection is therefore a quantitative phenomenon, but here again there are many interpretations. One possible explanation is that the reduced level of cortisone permits a low-level immunological attack by the host on the tapeworm, and that the parasite tolerates this attack by repairing its damaged tissues. A more probable explanation is that cortisone, given at the reduced dose of 0.5 mg thrice weekly, does not remain at a high enough level in the body to suppress attack throughout, and so, for a period immediately prior to the next dose, worm growth is impaired or stopped. The result would be a decrease in the weight of the worms until a new equilibrium was established between the smaller amount of new tissue being produced between cortisone doses and the loss of eggs and proglottids during that period. This latter explanation fits with the observation, quoted above, that 1.25 mg of cortisone every third day results in smaller worms than those found when it is given daily or every second day. There was no indication that tolerance had been induced by giving cortisone at the full effective level for the first 20 days PI.

In conclusion, cortisone is well tolerated by the host and is the most effective and most reliable way of preventing loss of *H. diminuta* (cf. Rose [1972], who wrote the same about coccidia). Its mode of action is difficult to determine, however, because of its vast array of potential effects (Baxter and Forsham, 1972; Claman, 1972, 1975). Nevertheless, in recent years much has been learned about the action of cortisone and this justifies further investigation of its function in preventing worm rejection. In particular, it would be interesting to discover what process is blocked so quickly following the injection of cortisone that worms which have destrobilated commence to grow again instead of being rejected by the host.

ACKNOWLEDGMENTS

This work was supported by grant no. G971/100 from the M.R.C. (London) and Fisons Pharmaceuticals, to both of whom we are most grateful. It is also a pleasure to thank Jack Keys for his technical assistance and Dean Befus for critically reading the manuscript.

REFERENCES CITED

Baggiolini, M.
    1972 The enzymes of the granules of polymorphonuclear leukocytes and their functions. Enzyme **13**:132-160.

Baxter, J. D. and P. H. Forsham
1972 Tissue effects of glucocorticoids. American Journal of Medicine 53:573-589.

Befus, A. D.
1974 The immunoglobulin coat on *Hymenolepis* spp. Transactions of the Royal Society of Tropical Medicine and Hygiene 68:273.

1975a Secondary infections of *Hymenolepis* diminuta in mice: effects of varying worm burdens in primary and secondary infections. Parasitology 71:61-75.

1975b Intestinal immune responses of mice to the tapeworms *Hymenolepis diminuta* and *H. microstoma*. Ph.D. dissertation, University of Glasgow.

Befus, A. D. and D. W. Featherston
1974 Delayed rejection of single *Hymenolepis diminuta* in primary infections of young mice. Parasitology 69:77-85.

Bellamy, J. E. C. and N. O. Nielsen
1974 Immune-mediated emigration of neutrophils into the lumen of the small intestine. Infection and Immunity 9:615-619.

Berenbaum, M. C.
1974 Comparison of the mechanisms of action of immunosuppressive agents. *In* Progress in Immunology II. L. Brent and J. Holborow, eds. New York: American Elsevier Publishing Company, Inc. Vol. 5, pp. 233-243.

Bienenstock, J.
1974 The physiology of the local immune response and the gastrointestinal tract. *In* Progress in Immunology II. L. Brent and J. Holborow, eds. New York: American Elsevier Publishing Company, Inc. Vol. 4, pp. 197-207.

Bland, P. W.
1976 Immunity to *Hymenolepis diminuta*: unresponsiveness of the athymic nude mouse to infection. Parasitology 72:93-97.

Claman, H. N.
1972 Corticosteroids and lymphoid cells. New England Journal of Medicine 287:388-397.

1975 How corticosteroids work. The Journal of Allergy and Clinical Immunology 55:145-151.

Coleman, R. M., J. M. Carty, and W. D. Graziadei
1968 Immunogenicity and phylogenetic relationship of tapeworm

antigens produced by *Hymenolepis nana* and *Hymenolepis diminuta*. Immunology **15**:297-304.

Dineen, J. K., J. D. Kelly, B. S. Goodrich, and I. D. Smith
1974 Expulsion of *Nippostrongylus brasiliensis* from the small intestine of the rat by prostaglandin-like factors from ram semen. International Archives of Allergy **46**:360-374.

Gewurz, H., P. R. Wernick, P. G. Quie, and R. A. Good
1965 Effects of hydrocortisone succinate on the complement system. Nature **208**:755-757.

Goodall, R. I.
1973 Studies on the growth, location specificity and immunobiology of some Hymenolepid tapeworms. Ph.D. dissertation, University of Glasgow.

Harris, W. G. and J. A. Turton
1973 Antibody response to tapeworm (*Hymenolepis diminuta*) in the rat. Nature **246**:521-522.

Hopkins, C. A., P. M. Grant, and H. Stallard
1973 The effect of oxyclozanide on *Hymenolepis microstoma* and *H. diminuta*. Parasitology **66**:355-365.

Hopkins, C. A. and H. E. Stallard
1974 Immunity to intestinal tapeworms: the rejection of *Hymenolepis citelli* by mice. Parasitology **69**:63-76.

Hopkins, C. A., G. Subramanian, and H. Stallard
1972a The development of *Hymenolepis diminuta* in primary and secondary infections in mice. Parasitology **64**:401-412.

1972b The effect of immunosuppressants on the development of *Hymenolepis diminuta* in mice. Parasitology **65**:111-120.

Jarrett, W. F. H., E. E. E. Jarrett, H. R. P. Miller, and G. M. Urquhart
1968 Quantitative studies on the mechanism of self-cure in *Nippostrongylus brasiliensis* infections. Proceedings of the Third International Conference of The World Association for the Advancement of Veterinary Parasitology, *in* The Reaction of the Host in Parasitism. E. J. L. Soulsby, ed. Marburg/Lahn: Elwert Universitäts und Verlagsbuchhandlung. Pp. 191-198.

Jones, V. E. and B. M. Ogilvie
1971 Protective immunity to *Nippostrongylus brasiliensis*. The sequence of events which expels worms from the rat intestine. Immunology **20**:549-561.

Keller, R. and R. Keist
1972 Protective immunity to *Nippostrongylus brasiliensis* in the rat. Central role of the lymphocyte in worm expulsion. Immunology 22:767-773.

Kelly, J. D., J. K. Dineen, and R. J. Love
1973 Expulsion of *Nippostrongylus brasiliensis* from the intestine of rats: evidence for a third component in the rejection mechanism. International Archives of Allergy 45:767-779.

Lewis, G. P. and P. J. Piper
1975 Inhibition of release of prostaglandins as an explanation of some of the actions of anti-inflammatory corticosteroids. Nature 254:308-311.

Moss, G. D.
1971 The nature of the immune response of the mouse to the bile duct cestode, *Hymenolepis microstoma*. Parasitology 62:285-294.
1972 The effect of cortisone acetate treatment on the growth of *Hymenolepis microstoma* in mice. Parasitology 64:311-320.

Rose, M. E.
1972 Immune response to intracellular parasites. II. Coccidia. *In* Immunity to Animal Parasites. E. J. L. Soulsby, ed. London: Academic Press.

Thomas, H. C. and D. M. V. Parrott
1974 The induction of tolerance to a soluble protein antigen by oral administration. Immunology 27:631-639.

Wakelin, D.
1970 Studies on the immunity of albino mice to *Trichuris muris*. Suppression of immunity by cortisone acetate. Parasitology 60:229-237.

Wakelin, D. and G. R. Selby
1974 The induction of immunological tolerance to the parasitic nematode *Trichuris muris* in cortisone-treated mice. Immunology 26:1-10.

Wardle, R. A. and N. K. Green
1941 The rate of growth of the tapeworm *Diphyllobothrium latum* (L). Canadian Journal of Research 19D:245-251.

Wassom, D. L., C. W. DeWitt, and A. W. Grundmann
1974 Immunity to *Hymenolepis citelli* by *Peromyscus maniculatus*: genetic control and ecological implications. The Journal of Parasitology 60:47-52.

Weinmann, C. F.

   1966 Immunity mechanisms in cestode infections. *In* Biology of Para-
        sites. E. J. L. Soulsby, ed. Philadelphia: Academic Press. Pp.
        301-320.

   1970 Cestodes and Acanthocephala. *In* Immunity to Parasitic Animals.
        G. J. Jackson, R. Herman, and I. Singer, eds. New York: Appleton.
        Vol. 2, pp. 1021-1059.

DENSITY DISTRIBUTION OF DNA
FROM PARASITIC HELMINTHS WITH SPECIAL
REFERENCE TO *ASCARIS LUMBRICOIDES*

by Araxie Kilejian and Austin J. MacInnis

ABSTRACT

DNA was isolated from eleven species of parasitic helminths including trematodes, cestodes, a nematode, and an acanthocephalan. The buoyant density and GC content for each was determined by analytical ultracentrifugation. These results indicated buoyant densities ranging from 1.720 to 1.697 gm/cc, corresponding to GC contents of 61% to 38%. Such results suggest the possibility that GC content (and hence AT content) may be correlated with the amount of exposure of life cycle forms to UV irradiation from sunlight, which induces T̂T dimers. Highest GC content was generally observed in those species possessing freeswimming larval stages.

Analysis of buoyant density satellite DNA was accomplished by fractionation of DNA using serial preparative centrifugation in CsCl followed by analytical centrifugation. Such studies on Hymenolepidid cestodes revealed satellites with similar buoyant densities in all members of the genus that were examined. Application of this technique to somatic and germ-line tissues of *Ascaris* demonstrated that DNA from both tissues could be resolved into four density components banding approximately at 1.690, 1.696, 1.700, and 1.710 gm/cc with the main nuclear peak at 1.700.

Quantitative differences in the proportion of these components were observed between germ-line and somatic tissue DNAs as well as that of eggs and sperm. In egg DNA the peak at 1.690 was of greater magnitude that the main nuclear peak, while in sperm samples it appeared only as a minor satellite, similar to the somatic tissue DNAs. Electron microscopy of this egg satellite showed mainly circular molecules, indicating its mitochondrial origin. DNA from both egg and sperm showed an augmented peak at 1.696 in comparison to somatic tissue DNAs. In the

Araxie Kilejian is Associate Professor of Parasitology at The Rockefeller University. Austin MacInnis is Professor of Biology at the University of California, Los Angeles.

latter, this component was apparent only after fractionation. These results suggest a substantial loss of this density component during chromatin elimination, without excluding the possibility of elimination from other components as well.

INTRODUCTION

Studies of DNA from a wide range of organisms have shown that species which are closely related taxonomically have similar DNA base compositions. The observation that base compositions $(G + C/A + T)$ can be calculated from buoyant densities of DNAs (with the exception of those containing rare bases) has enabled such analyses on small quantities of DNA (Schildkraut et al., 1962). Information on the density distribution of DNA from parasitic helminths is limited to *Ascaris lumbricoides* (Bielka et al., 1968; Ward, 1971; Tobler et al., 1972); *Hymenolepis diminuta* (Carter et al., 1972; Carter and MacInnis, in preparation), and *Schistosoma mansoni* (Hillyer, 1974). In the course of our studies on various properties of DNA from parasitic helminths, we measured the buoyant densities of DNAs isolated from ten additional species and observed density satellites in most samples. Satellites are often not evident by analytical pycnography of total DNA and can be resolved only after fractionation of the DNA on preparative CsCl density gradients (McConaughy and McCarthy, 1970). Adequate amounts of test materials were not available from all species; therefore, complete analyses of satellite DNAs were limited to *H. citelli*, *H. microstoma*, and *A. lumbricoides*. The density satellites of *Ascaris* DNA prepared from germ-line as well as somatic tissue were studied in detail to determine whether chromatin elimination during development (Meyer, 1895; Bonnevie, 1902) involves a DNA component with a specific density. Bielka et al. (1968) had shown that DNA from fertilized eggs of *A. lumbricoides* banded at a density of 1.697 g/cc, with a minor satellite at 1.693 g/cc and a major one at 1.685 g/cc. DNA from the gastrula showed considerable reduction in the satellite at 1.685 g/cc. Bielka and colleagues did not correlate this reduction with nuclear chromatin elimination, but proposed a cytoplasmic origin, possibly mitochondrial, for the light satellite. However, the studies of Carter et al. (1972) showed that purified mitochondrial DNA from *Ascaris* testes is circular and bands at a density of 1.690 g/cc. Two additional conflicting reports on *Ascaris* DNA have been published (Ward, 1971; Tobler et al., 1972). These are discussed with our findings.

MATERIALS AND METHODS

Hymenolepidids and *Moniliformis dubius* were maintained in the laboratory by established methods (see MacInnis and Voge, 1970).

Adult *Ascaris lumbricoides* were collected live from the small intestines of pigs at a Los Angeles abattoir and on arrival at the laboratory placed in 0.85% saline at 37° C for maintenance. All tissues were isolated within 4 hours of collection. After removal of a worm's viscera, the edge of a microscope slide was used to strip the muscles from the cuticle. To collect sperm, seminal vesicles were carefully removed, held on the side of a centrifuge tube with forceps and the contents drained by making a small incision on the vesicle wall. The tube contents were centrifuged and the resulting pellet was used to extract DNA. Excised *Ascaris* intestines were flushed clean by forcing 30-40 mls of saline through them with a hypodermic needle and syringe. For the extraction of DNA from the uterine wall, the terminal 3-4 cm of the uteri were used. Fertilized eggs were collected by the method of Costello (1961), and homogenized by a French pressure cell at 5000-7000 psi.

Ethanol-fixed samples of *Gyrocotyle rugosa* (from rat fish, Friday Harbor, Washington) were donated by Dr. John Simmons. *Lacystorhynchus*, *Orygmatobothrium* and *Phyllobothrium* (from *Mustelis canis*, Bodega Bay, California) were collected with the aid of Drs. Clayton Page and John Simmons. Lyophilized *Schistocephalus* (from sticklebacks, Glasgow, Scotland) was a gift from Adrian Hopkins. Lyophilized *Echinococcus multilocilaris* brood capsules (laboratory reared in cotton rats) were a gift from Dr. Dan Harlow. Ethanol-fixed *Fasciola hepatica* (from cattle, Houston, Texas) were collected with aid from Drs. Glen Harrington and John Simpson.

The DNA of these various tissues was extracted and purified by a combination of published methods. All collected samples were macerated or homogenized in a NET buffer (0.5 M NaCl, 0.1 M EDTA and 1.15 M Tris, pH 8.5). Sodium dodecyl sulfate was added to a concentration of 10 mg/ml and the samples were heated (10 minutes, 60° C). They were transferred to a 37° C water bath and pronase (preincubated: 37° C, 30 minutes) was added to give a concentration of 1 mg/ml. Following gentle shaking overnight in pronase, the samples were deproteinized once using chloroform-isoamyl alcohol (24:1 v/v) and dialyzed in 2 × SSC (0.3 M NaCl, 0.03 M sodium citrate, pH 7.5). Samples were concentrated by dialysis, then treated with α-amylase ($100\mu g$/ml) and pancreatic ribonuclease A and T_1, ($100\mu g$/ml and 100 units/ml, respectively). Further purification as well as fractionation of total DNA according to buoyant density was accomplished by preparative CsCl equilibrium centrifugation. Using a preparative ultracentrifuge (Spinco Model L2) and a fixed angle 50 rotor, 200-400 μg DNA was centrifuged at 33000 rpm for 60 hours. Four-drop fractions were collected from the bottom of the gradient and diluted with water, and their absorbancies at 260 nm were measured. Appropriate fractions across the gradient were pooled and rerun in a Spinco Model E analytical ultracentrifuge (20 hours, 44000 rpm at 25° C). Buoyant densities and GC (guanine + cytosine) content were calculated by the method of Schildkraut et al. (1962), *Micrococcus lysodeikticus* DNA being used as a density marker (1.731 gm/cc). Our previous studies (Simmons

164   RICE UNIVERSITY STUDIES

et al., 1972) demonstrated that alcohol preserved material was adequate for such studies. All chemicals used were reagent grade. Pronase, α-amylase, and ribonuclease (T_1, and pancreatic 5$\times$ Cryst) were purchased from Calbiochem.

RESULTS

GC Content

Buoyant densities of the main components of isolated DNAs and their base composition calculated from these densities are summarized in table 1. Despite the phylogenetic diversity of helminth groups studied, with the exception of *Orygmatobothrium*, no extreme differences were observed. However, a relatively higher GC content of *Orygmatobothrium*, *Schistocephalus*, *Fasciola*, *Gyrocotyle*, and *Lacystorhynchus* is apparent as compared to *Moniliformis*, *Ascaris*, and the three hymenolepidids. Singer and Ames (1970) proposed that bacterial species naturally exposed to sunlight have evolved a high GC content (and thus low adenine + thymine) as a means of avoiding genetic damage from thymine dimers formed by solar radiation. Although insufficient evidence is so far available to apply this speculation to higher organisms in general, our data appear to suggest that parasites having a free-swimming larval stage exposed to sunlight will have a relatively higher GC content of their DNA. *Echinococcus multilocularis* and the Schistosomes

TABLE I

ORGANISM	BUOYANT DENSITY OF MAIN BAND	% G + C
Orygmatobothrium sp.*	1.720	61
*Fasciola hepatica**	1.706	47
Schistocephalus sp.**	1.706	47
*Gyrocotyle rugosa**	1.705	46
*Lacystorhynchus tenuis**	1.704	45
Phyllobothrium sp.*	1.703	44
*Echinococcus multilocularis***	1.703	44
Ascaris lumbricoides	1.700	41
Moniliformis dubius	1.698	39
Hymenolepis microstoma	1.698	39
H. citelli	1.698	39
H. diminuta	1.697	38
*Schistosoma****	1.693	34

* from tissue fixed in ethanol
** lyophilized tissue; all other tissue was fresh
*** from Hillyer, 1974

appear as a possible exception. Such speculations, however, must eventually be correlated with presence or absence of the enzymes associated with repair of DNA lesions. The calculations of GC content from buoyant density must be viewed with caution, since we have no data for some species to preclude the possibility of presence of rare bases.

Analysis of Satellites: Cestodes

The microdensitometer tracings of most DNAs centrifuged to equilibrium in CsCl revealed satellite peaks. While the function of satellite DNAs observed in several organisms remains unknown, some have been shown to represent mitochondrial DNA or code for ribosomal RNA. Carter and MacInnis (unpublished) have analyzed the satellite DNAs of *H. diminuta* in some detail. In addition to the main nuclear band at 1.696 g/cc, they have observed four satellites with densities of 1.691, 1.705, 1.708, and 1.717 g/cc, respectively. The satellite at 1.691 g/cc was shown to be mitochondrial DNA (Carter et al., 1972). To compare the finding on *H. diminuta* with two closely related species, total DNAs from *H. citelli* and *H. microstoma* were each fractionated into light and heavy density components on preparative CsCl gradients. The density distribution of DNA from these fractions is shown in figures 1 and 2. The light fraction of *H. microstoma* revealed a satellite at 1.692 g/cc (similar to the DNA from *H. diminuta*). Our inability to demonstrate this peak in the light fraction of *H. citelli* does not preclude its presence. The small quantity of this satellite could be masked easily by the presence of much greater quantities of nuclear DNA, or it could have been lost in preparation of the sample. In the heavy fractions of *H. citelli* and *H. microstoma* in addition to the three satellites similar in density to those reported for *H. diminuta*, there is also a peak at 1.722 g/cc. It would be of interest to determine whether this fourth satellite is indeed absent in *H. diminuta*.

Satellites and Chromosome Diminution in *Ascaris*

Since the fractionating procedure of total DNA from the hymenolepidid species showed good resolution of satellite peaks, it was applied to DNA isolated from different tissues of *Ascaris* to determine whether chromatin eliminated from somatic cells during early cleavage can be identified as a distinct satellite.

Fertilized eggs and sperm were used for the isolation of DNA before chromatin diminution; intestinal wall, uterus, and muscle were used for samples subsequent to diminution. Analytical CsCl density gradient patterns of total DNAs from all these tissues indicated the presence of more than a single DNA component in each (figure 3). While DNAs from somatic tissues (figure 3a,b,c) gave a main band at density ca. 1.700 g/cc and only a shoulder at 1.690 g/cc, that from the egg was strikingly different (figure 3e). The major component of egg DNA banded at density 1.690 g/cc. In addition, there was a peak at 1.700 g/cc, a distinct component at 1.696 g/cc, and a heavy satellite

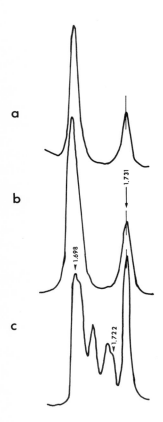

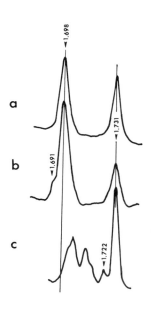

FIG. 1. ANALYTICAL PYCNOGRAPHY OF *HYMENOLEPIS CITELLI* DNA. a, total DNA; b, aliquot of pooled fractions from the light side of peak shown in a; c, aliquot of fractions from the heavy side of peak shown in a. Tracings were superimposed with the marker DNAs aligned.

FIG. 2. ANALYTICAL PYCNOGRAPHY OF *HYMENOLEPIS MICROSTOMA* DNA. a, total DNA; b, aliquot of pooled fractions from the light side of peak shown in a; c, aliquot of fractions from the heavy side of peak shown in a. Tracings were superimposed with the marker DNA aligned.

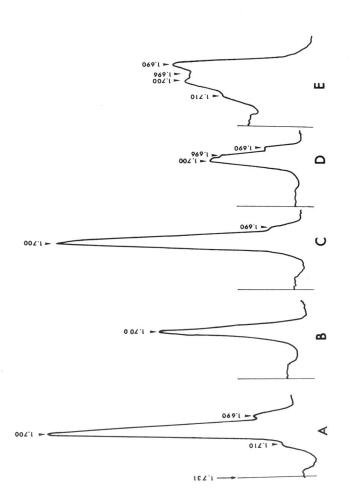

FIG. 3. MICRODENSITOMETER TRACINGS OF UV ABSORBANCE PATTERNS of *Ascaris* DNAs centrifuged to equilibrium in CsCl in a Spinco Model E analytical ultracentrifuge at 44,000 rpm for 20 hours. a, muscle DNA; b, uterine wall DNA; c, intestinal wall DNA; d, sperm DNA; e, fertilized egg DNA. Marker DNA with a density of 1.731 (arrow) was added to all samples. The line on left of each tracing indicates position of marker.

band at 1.710 g/cc. A similar heavy satellite was also evident in muscle (figure 3a). The density distribution pattern of sperm DNA was of special interest (figure 3d). Since the germinal cells supposedly arise from cell lines that do not undergo chromatin diminution, it was expected that sperm DNA would reflect a complete, undiminished DNA pattern similar to that of fertilized eggs. Unlike somatic tissue DNA, sperm samples did show the presence of a peak at 1.696 g/cc, as was also seen in the egg. However, the DNA band at 1.690 g/cc, prominent in the egg, was by comparison considerably smaller.

From the above results it could be concluded that (1) the peak at 1.700 g/cc represents the major nuclear DNA component that persists after chromatin elimination; (2) the major egg DNA peak at 1.690 g/cc is less prominent in sperm as well as somatic tissue DNAs; (3) the only apparent qualitative difference between germinal and somatic tissue DNAs is the absence of the density component at 1.696 g/cc from the latter.

Fractionated egg DNA (figure 4) confirmed the presence of four distinct density peaks as already seen in total DNA samples. The difficulty of collecting sufficient quantities of sperm limited a detailed fractionation of the DNA. However, separation of the total DNA sample into a light and heavy fraction did reveal distinct density peaks of 1.690, 1.696, and 1.702 g/cc (figure 5). Fractionation of DNA from the uterine wall gave unexpected results. What had appeared as a fairly symmetrical peak in total DNA samples (figure 3b) could now be resolved into four distinct bands (figure 6) with the same buoyant densities as seen in egg DNA. The banding pattern of total DNA from intestinal wall had clearly shown a light satellite at a density of 1.690 in addition to the main peak at 1.700 g/cc (figure 3c). Additional peaks could not be resolved in an initial attempt at fractionation (figure 7a,b,c). When the light fraction (figure 7a) was refractionated, however, a clear peak at 1.696 g/cc became evident (figure 7d). These results of fractionated uterine and intestinal wall DNAs illustrate clearly that what had appeared as a qualitative difference between total DNAs of germinal and somatic tissues is only a quantitative difference.

It would be of interest to determine whether this component at 1.696 g/cc is also present in muscle DNA. The presence of large quantities of glycogen as well as some UV-absorbing contaminants made isolation of the relatively minute quantities of pure muscle DNA a major task. A single successful preparation showed a perfectly symmetrical main band at 1.699 g/cc and two satellite bands at 1.690 and 1.710.

Discussion

We do not consider the absence of the density satellite of 1.710 g/cc from sperm and intestinal wall DNA to be significant. This could have resulted from our failure to collect enough of the heavy side of the CsCl gradients that did not show detectable absorption at 260 nm.

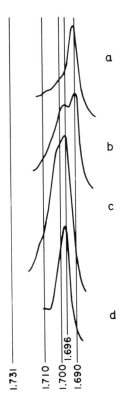

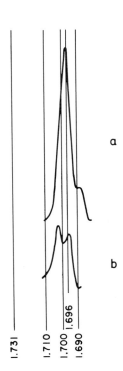

FIG. 4. ANALYTICAL PYCNOGRAPHY OF *AS-CARIS* EGG DNA. About 300 μg of total native DNA was centrifuged to equilibrium in CsCl at 33,000 rpm in a fixed angle 50 rotor for 60 hours. Four-drop fractions were collected from the bottom of the tubes. Aliquots of selected, pooled fractions across the peak were then centrifuged to equilibrium in CsCl in the analytical ultracentrifuge. The tracings are arranged in sequence (a-d) starting with the lowest density fraction selected, and were superimposed with the marker DNAs aligned. Other conditions as in figure 3.

FIG. 5. ANALYTICAL PYCNOGRAPHY OF *ASCARIS* SPERM DNA. Conditions as in figure 4.

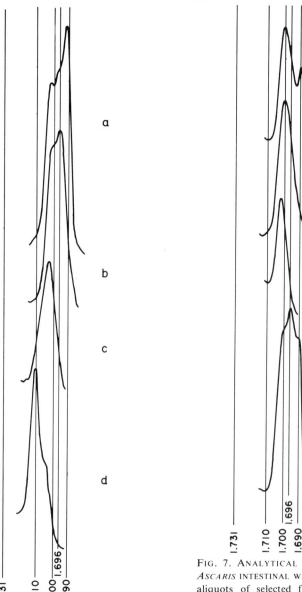

FIG. 6. ANALYTICAL PYCNOGRAPHY OF *ASCARIS* UTERINE WALL DNA. Conditions as in figure 4.

FIG. 7. ANALYTICAL PYCNOGRAPHY OF *ASCARIS* INTESTINAL WALL DNA. a-c are aliquots of selected fractions from the initial, preparative CsCl equilibrium centrifugation. 100 μg of fraction b was centrifuged to equilibrium in the preparative ultracentrifuge, and d is a light fraction from this peak analyzed with the Model E. Other conditions as in figure 4.

Because of differences of density values reported by different investigators, comparison of results from different labs becomes difficult. Small variations in buoyant densities arise from altered relative proportions of adjacent peaks in particular fractions tested (clearly demonstrated in our fractionated samples), from slight changes of the banding position of different samples within CsCl gradients, or from overloading samples (as we have often done in searches for satellite bands). We have shown that the DNA from both germinal and somatic tissues of *A. lumbricoides* can be resolved into four density components banding at 1.689-1.691, 1.694-1.696, 1.699-1.702, and 1.710-1.711 g/cc. To facilitate discussion we will refer to them by the most prevalent values obtained, i.e., density peaks of 1.690, 1.696, 1.700, and 1.710 g/cc.

Initially our results appeared different from those reported by Bielka et al. (1968); if their density values were shifted by .003 units to equate the main nuclear band values at 1.700 g/cc, however, then their egg satellite peak at 1.693 g/cc would coincide with the 1.696 g/cc band of our samples and that at 1.685 g/cc would be close to our value of 1.690 g/cc. Also, the reduction of the 1.693 and 1.685 g/cc peaks in their tracings of DNA from gastrula would be consistent with our findings that density peaks at 1.696 and 1.690 g/cc are reduced in somatic tissue DNAs. Carter et al. (1972) showed that the DNA of purified mitochondria isolated from *Ascaris* testes was circular and banded at a density of 1.690 g/cc. Electron micrographs of the egg fraction shown in figure 3a showed circular molecules in addition to linear fragments. The mitochondrial origin of this satellite would explain the reduction of this peak in sperm DNA as compared to fertilized egg DNA. Eggs of many species have been shown to have an abundance of mitochondrial DNA (Davidson, 1968); also, during fertilization the total contents of *Ascaris* sperm (including mitochondria) fuse with the oocyte cytoplasm (Foor, 1967). However, Ward (1971) reported that DNA of fertilized *Ascaris* eggs shows only two components with densities of 1.692 and 1.701 g/cc. He proposed that the light band, which is 52% of the egg DNA, represents the chromatin eliminated during cleavage, since it is absent from the gastrula and the eggs contain RNA which hybridized specifically with this light satellite. His conclusion of a nuclear origin for the light satellite would imply that our 1.690 g/cc band is a mixture of mitochondrial and nuclear DNA. We have not been able to isolate clean nuclei from *Ascaris* eggs to preclude this possibility, but our electron micrographs of this fraction indicate clearly that mitochondrial DNA contributes to the peak. In another recent publication, Tobler et al. (1972) compared DNAs isolated from *A. lumbricoides* spermatids, 4-cell stage developing eggs, and larvae. They observed a single peak at 1.697 g/cc in larval and spermatid DNA and concluded that mitochondrial DNA contributes little to spermatid DNA. DNA prepared from 4-cell stages showed an additional satellite band at 1.686 g/cc, which they consider to be of mitochondrial origin since electron microscopy showed circular molecules. From these observations, they concluded that eliminated DNA does not differ in density from somatic or germ-line DNA. These investigators have used ethanol precipitation and spooling during the isolation procedure, however, and report only 40% recovery of DNA using the isotope dilution method. This procedure could result in a possible differential loss of some DNA fractions, as is evident in the loss of mitochondrial DNA from their preparation of spermatid DNA. Also, they have assumed that chromatin diminution in the strain of *Ascaris* they studied does not take place until the third cleavage, following the observations of Meyer and Bonnevie made in 1895 and 1902. Opinion differs as to whether chromatin elimination starts at second or third cleavage for *Ascaris equorum* (Fogg, 1930).

Obviously, the nature of the eliminated chromatin during the cleavage of *Ascaris lumbricoides* eggs still remains a problem requiring further study. We have shown clearly an identical heterogeneity of density components within germinal and somatic tissue DNAs with only some quantitative differences. An overall reduction of chromatin from all density bands cannot be precluded with our experimental methods; however, our findings indicate that of the four DNA components the most obvious reduction is from the peak at 1.696 g/cc. The proportion of mitochondrial components to a possible nuclear DNA component at 1.690 g/cc needs reinvestigation.

ACKNOWLEDGMENTS

This work was supported by Grants NSF GB 8753, NIH-AI 00070, University of California Research 2280, and in part by the UCLA Life Sciences General Support Grant. The technical assistance of Mrs. June Baumer is gratefully acknowledged.

REFERENCES CITED

Bielka, H., I. Schultz, and M. Böttger
1968 Isolation and properties of DNA from eggs and gastrulae of *Ascaris lumbricoides*. Biochimica et Biophysica Acta **157**:209-212.

Bonnevie, K.
1902 Ueber Chromatindiminution bei Nematoden. Zeitschrift für Naturwissen **36**:275-288.

Carter, C. E., J. R. Wells, and A. J. MacInnis
1972 DNA from anaerobic adult *Ascaris lumbricoides* and *Hymenolepis diminuta* mitochondria isolated by zonal centrifugation. Biochimica et Biophysica Acta **262**:135-144.

Costello, L. C.
  1961 A simplified method of isolating *Ascaris* eggs. Journal of Para-
         sitology **47**:24.

Davidson, E. H.
  1968 Gene Activity in Early Development. Academic Press, New York.
         Pp. 244-246.

Fogg, L. C.
  1930 A Study of Chromatin Diminution in *Ascaris* and *Ephestia*.
         Journal of Morphology and Physiology **50**:413-450.

Foor, W. E.
  1967 Ultrastructural aspects of oocyte development and shell formation
         in *Ascaris lumbricoides*. Journal of Parasitology **53**:1245-1261.

Hillyer, G. V.
  1974 Buoyant density and thermal denaturation profiles of Schistosome
         DNA. Journal of Parasitology **60**:725-727.

MacInnis, A. J. and M. Voge
  1970 Experiments and Techniques in Parasitology. San Francisco:
         W. H. Freeman and Co.

McConaughy, B. L. and B. J. McCarthy
  1970 Related base sequences in the DNA of simple and complex organ-
         isms. VI. The extent of base sequence divergence among the DNAs
         of various rodents. Biochemical Genetics **4**:425-446.

Meyer, O.
  1895 Cellulare Untersuchungen an Nematoden-Eiern. Zeitschrift für
         Naturwissen **29**:391-410.

Schildkraut, C. L., J. Marmur, and P. Doty
  1962 Determination of the base composition of deoxyribonucleic acid
         from its buoyant density in CsCl. Journal of Molecular Biology
         4:430-443.

Simmons, J. E., G. H. Buteau, Jr., A. J. MacInnis, and Araxie Kilejian
  1972 Characterization and Hybridization of DNAs of Gyrocotylidean
         parasites of Chimaeroid fishes. International Journal for Para-
         sitology **2**:273-278.

Singer, C. E. and B. N. Ames
  1970 Sunlight ultraviolet and bacterial DNA Base Ratios. Science
         170:822-826.

Tobler, H., K. D. Smith, and H. Ursprung
1972 Molecular aspects of chromatin elimination in *Ascaris lumbricoides*. Developmental Biology **27**:190-203.

Ward, K. A.
1971 Satellite DNA in relation to chromatin diminution in developing eggs of *Ascaris lumbricoides*. Dissert. Abstr. Int. B, 636.

ULTRASTRUCTURAL CHANGES IN THE INFECTIVE LARVAE OF *NIPPOSTRONGYLUS BRASILIENSIS* IN THE SKIN OF IMMUNE MICE

by D. L. Lee

ABSTRACT

Infective stage larvae of *Nippostrongylus brasiliensis* are immobilized within two to three hours after penetrating the skin of mice that are immune to this nematode. The larvae become surrounded by host defense cells and bundles of collagen fibers. The cuticle is the first structure of the larva to be attacked; host defense cells may secrete a collagenase which attacks the cuticle. Disorganization of the hypodermis and underlying muscle cells follows destruction of the larval cuticle.

INTRODUCTION

Infective third-stage larvae of *Nippostrongylus brasiliensis* quickly enter the skin of rats and mice, and move freely through the dermis. Larvae in the skin appear to be more permeable to fixative than those on the surface of the skin; this difference may be related to the change in environment and temperature when the nematode enters the skin of the mammalian host. In primary infections, host cells seem not to attack the penetrating larvae (Lee, 1972a). Taliaferro and Sarles (1937) have shown that larvae which penetrate the skin of immune hosts become coiled and immobilized in the skin or lungs, disintegrate and, eventually, are phagocytosed. This paper describes the ultrastructure of infective larvae in the skin of immunized mice.

MATERIALS AND METHODS

Four mice, six weeks of age, were each exposed to 500 infective-stage larvae of *Nippostrongylus brasiliensis*. Hair was clipped from an area on the left side of the abdomen of each mouse, and the larvae were applied to this area in a drop of water. The mice were physically restrained until the water had evaporated (five minutes). The initial exposure to 500 larvae was repeated two and

---

D. L. Lee is Professor of Agricultural Zoology at the University of Leeds.

four weeks later. Six weeks after the initial exposure to the larvae each mouse was exposed to 1000 larvae. These larvae were applied as before, but to an area on the right side of the abdomen. The larvae were allowed to penetrate for two or three hours. After the mice had been killed, the area of skin containing larvae on the right side of the abdomen was removed, chopped into pieces in cold 2.5% gultaraldehyde in cacodylate buffer (pH 7.0), and fixed at 4° C for 24 hours, or chopped into pieces in 1% osmium tetroxide (Rosenbluth, 1965) and fixed for 2 hours at 4°C. The fixed tissue was washed in buffer, dehydrated in ethanol, transferred to propylene oxide and then embedded in Araldite. Sections were cut on a Huxley ultramicrotome or an LKB Ultrotome III, mounted on formvar-coated grids, stained in a 5% solution of uranyl acetate in methanol followed by lead citrate, and viewed by means of an AEI6B or a Philips EM300 electron microscope.

RESULTS

Larvae were present in the dermis two to three hours after application to the skin, and most of them were coiled. Initial observations suggested that the worms were immobilized but were otherwise normal, as normal structural features such as esophagus, intestine, and muscles were readily apparent. Closer investigation, however, revealed a number of significant alterations in

---

ABBREVIATIONS FOR FIGURES

c, degenerating cuticle of body wall; co, collagen fibers of host origin; d, dermis of host skin; h, hypodermis; m, mitochondria; mu, muscle of body wall; my, myofilaments; o, esophagus; oc, esophageal cuticle.

FIG. 1. ELECTRON MICROGRAPH OF A SECTION THROUGH A COILED UP, INFECTIVE-STAGE LARVA of N. brasiliensis in the skin of an immune mouse. The larva is sectioned in three places with the outermost part of the coil in the top of the picture and the parts of the worm in the center of the coil in the lower half of the picture. The cuticle has completely disappeared where the larva is exposed to host tissue but is still present, although badly damaged, in the more centrally placed regions of the coil. × 11,000.

FIG. 2. ELECTRON MICROGRAPH OF A SECTION THROUGH THE ESOPHAGUS of an infective-stage larva of N. brasiliensis in the skin of an immune mouse. Note the almost normal appearance of the esophagus and its cuticle. × 13,000.

FIG. 3. ELECTRON MICROGRAPH OF A SECTION THROUGH THE BODY WALL of an infective-stage larva of N. brasiliensis in the skin of an immune mouse. Note the disorganized appearance of the myofilaments, the collagen fibers between the muscles and the hypodermis, the absence of cuticle, and the presence of collagen fibers on the outer surface of the hypodermis. × 27,000.

FIG. 4. ELECTRON MICROGRAPH OF A SECTION THROUGH THE BODY WALL of an infective larva of N. brasiliensis in the skin of an immune mouse. Note the absence of the hypodermis, the slightly disorganized structure of the muscles, the almost normal appearance of the mitochondria, and the collagen fibers around the worm. × 23,000.

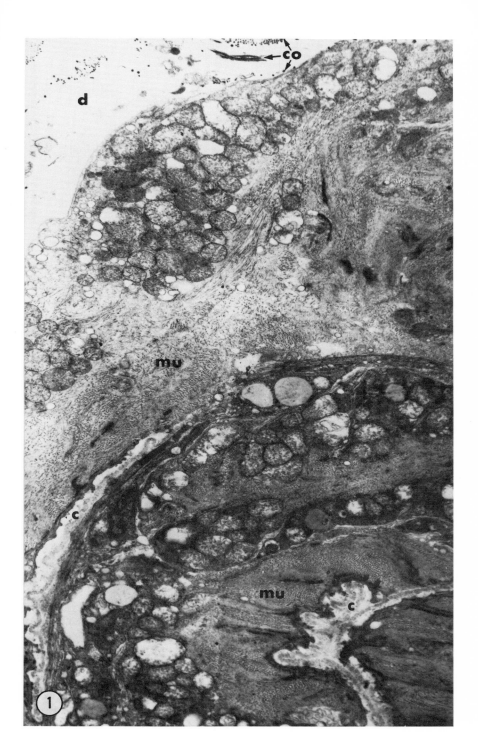

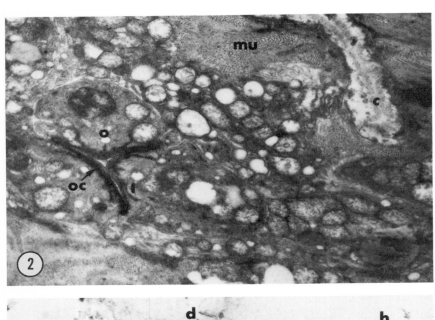

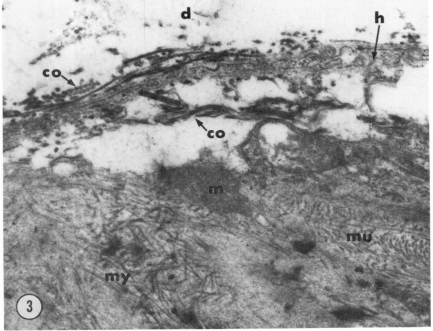

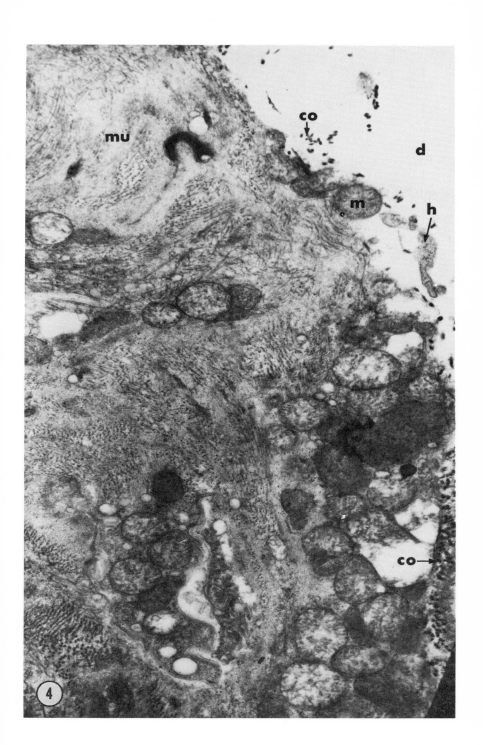

the structure of these larvae. They were usually surrounded by bundles of collagen fibers, and host cells (such as macrophages and fibrocytes) lay among the collagen fibers. Many larvae had no cuticle, or had remnants of an obviously degenerated cuticle (figures 1-4), and fibers of collagen, apparently of host origin, were in the position normally occupied by the cuticle. The cuticle on the outside of tightly coiled worms had disappeared, and although the cuticle on the central parts of the coil was present, it had degenerated (figures 1-3). The thin hypodermis was still present in some worms, but fibers of collagen were interspersed between it and the muscles of the body wall. In others, the hypodermis had partly or entirely disappeared, with the exception of the hypodermal cords. The muscles in the more centrally placed portions of the coiled larvae were fairly normal in appearance, but the muscles towards the periphery of these worms showed varying degrees of disorganization. In the worst affected muscles the myofilaments had lost their regular arrangement and were lying in all directions within the muscle cell, although thick and thin myofilaments were still easily recognizable. Most of the mitochondria in these disorganized larval muscle cells were normal in appearance, although a few were distended. The muscle nuclei appeared normal. The esophagus, including the esophageal cuticle, and the intestine of coiled worms were apparently unaltered in structure at this time.

DISCUSSION

Recent work on the penetration of host skin by larval stages of parasitic nematodes has concentrated on the structure and behavior of these larvae as they penetrate the skin of previously uninfected animals (Lee, 1972a; Matthews, 1972). Sarles and Taliaferro (1936) showed that in rats which were actively immune to *N. brasiliensis*, the infective larvae of this same nematode were prevented from completing their normal pattern of migration within the host. In the immunized animal, larvae were killed in the skin or lungs, or were prevented from growing, establishing themselves, or laying eggs in the intestine. Taliaferro and Sarles (1937 and 1939) later showed that the larvae become coiled and immobilized in the skin within the first few hours after penetration. They suggested that this occurs because of the presence of antibody. Precipitates form in and around coiled larvae, and an intense cellular infiltration (involving host eosinophils, macrophages, lymphocytes, monocytes, and heterophils) develops in the tissues surrounding the larvae. After the worm dies it is gradually digested by the surrounding macrophages.

Ogilvie (1965) has shown that the adult worm residing in the host intestine induces greater immunity than the migrating larvae.

The work described in this paper shows that migrating larvae in the skin of immune mice are immobilized within two or three hours after application to the skin. Presumably, the immobilization is brought about by antibodies induced by previous experience with larval and adult antigens.The larval cuticle is the first structure to be attacked by the immune host. Since the cuticle of most nematodes is a highly resistant collagenous structure, this finding is of considerable interest (see Lee, 1966 and 1972b). These results suggest that the cuticle of the immobilized larva is rapidly attacked by a collagenase that is, presumably, secreted by host defense cells surrounding the worm. The fact that the body wall cuticle covering the inner coils of the worm or located in the center of the esophagus is partially protected from degradation suggests that the observed effect is not the result of autolysis. The cytoplasmic parts of the nematode remain in a recognizable form for much longer than the cuticle, and only begin to break down once the cuticle has been destroyed. The coiled larva is quickly surrounded by collagen fibers of host origin. Following destruction of the cuticle, these fibers surround the hypodermis and eventually impose themselves between other cells of the nematode body wall. This sequence of events appears to be the beginning of the walling-off process that forms the nodule described by Taliaferro and Sarles (1937 and 1939). The speed with which the process occurs suggests that it is very efficient in protecting mice against secondary infections of *N. brasiliensis.*

REFERENCES CITED

Lee, D. L.
  1966 The structure and composition of the helminth cuticle. Advances in Parasitology **4**:187-254.

  1972a Penetration of mammalian skin by the infective larva of *Nippostrongylus brasiliensis*. Parasitology **65**:499-505.

  1972b The structure of the helminth cuticle. Advances in Parasitology **10**:347-379.

Matthews, B. E.
  1972 Invasion of skin by larvae of the cat hookworm, *Ancylostoma tubaeforme*. Parasitology **65**:457-467.

Ogilvie, B. M.
  1965 Role of adult worms in immunity of rats to *Nippostrongylus brasiliensis*. Parasitology **55**:325-335.

Rosenbluth, Jack
  1965 Ultrastructural organization of obliquely striated muscle fibers in *Ascaris lumbricoides*. Journal of Cell Biology **25**(3):495-515.

Sarles, M. P. and W. H. Taliaferro
  1936 The local points of defense and the passive transfer of acquired

immunity to *Nippostrongylus muris* in rats. Journal of Infectious Diseases **59**:207-220.

Taliaferro, W. H. and M. P. Sarles

1937 The mechanism of immunity to *Nippostrongylus muris*, the intestinal nematode of the rat. Science **85**:49-50.

1939 The cellular reactions in the skin, lungs and intestine of normal and immune rats after infection with *Nippostrongylus muris*. Journal of Infectious Diseases **64**:157-192.

SPECIFICITY OF AMINO ACID TRANSPORT IN THE TAPEWORM *HYMENOLEPIS DIMINUTA* AND ITS RAT HOST

by *A. J. MacInnis, D. J. Graff, A. Kilejian, and C. P. Read*

ABSTRACT

The specificity of amino acid transport loci in *Hymenolepis diminuta* was determined by assaying reciprocal inhibitions of uptake in pairs of all possible combinations of fifteen amino acids with the exception of glutamic acid and tyrosine, where metabolism and solubility, respectively, prevented acquisition of data. From these results we propose that *H. diminuta* possesses at least six distinct loci, but with overlapping affinities. These loci were designated as follows: a) dicarboxylic amino acid preferring; b) serine preferring (*A* site in other organisms); c) leucine preferring (*L* site in other organisms); d) aromatic amino acid preferring; e) dibasic amino acid preferring; f) glycine preferring. Methionine shows high affinity for all loci except the dibasic amino acid preferring locus. Histidine uses both the aromatic and dibasic loci. When the various amino acids were used in the external medium to stimulate efflux, the results corroborated the loci proposed from the inhibitor studies. Fifteen amino acids were each studied as inhibitors of the uptake of each of ten amino acids by segments of the rat's intestine. These results indicated the presence of a locus with high affinity for the dibasic amino acids in the rat's intestinal mucosa, but lysine clearly has overlapping affinity for the systems transporting the neutral amino acids. The major difference observed between the tapeworm and its host was the strong inhibition of alanine uptake by leucine in the rat gut, whereas leucine had less effect on alanine uptake by the worm.

Austin MacInnis is Professor of Biology at the University of California, Los Angeles. D. J. Graff is Associate Professor of Zoology at Weber State College, Ogden, Utah. A. Kilejian is Associate Professor of Parasitology at The Rockefeller University. C. P. Read is deceased.

INTRODUCTION

Over the last ten years, a considerable body of data has accumulated which indicates that the mediated transport of amino acids into and out of cells involves several systems, some of which show overlapping affinities for individual amino acids. These affinities seem to differ from one cell type to another and from one organism to another, and may differ in various organs. Read et al. (1963) concluded that in the tapeworm *Hymenolepis diminuta* there are at least four membrane mechanisms for amino acid transport. These show respective preferences for diamino, dicarboxylic, and two groups of monoamino-monocarboxylic acids; the systems do not show complete specificity for the indicated classes of amino acids. The same authors also presented evidence suggesting that the two mechanisms (loci) for the transport of "neutral" amino acids seemed to resemble the *A* (alanine-preferring) and *L* (leucine-preferring) loci reported for Ehrlich ascites cells (Oxender and Christensen, 1963); these loci were later found in the human erythrocyte (Winters and Christensen, 1964).

Read et al. (1963) postulated that the absorption of a single amino acid from an amino acid mixture would depend on the relative concentrations of other amino acids and the effectiveness of individual amino acids as competitive inhibitors of adsorption of the single amino acid concerned. Experimental tests of this hypothesis have supported it in the cases of two tapeworm species (Read et al., 1963; Senturia, 1964) and two acanthocephalan species (Rothman and Fisher, 1964).

Transport loci's differing but overlapping specificities for amino acids may be presumed to involve (a) regulation of the intracellular pools of free amino acids in the intestinal parasite and in the host intestinal mucosa; (b) regulation of the composition of the extracellular pool of free amino acids in the gut lumen; and (c) the level of competition between host and parasite for available amino acids.

Since we still lack comprehensive data on the specificity of amino acid transport loci in an animal parasite, we have made a systematic examination of the transport interactions of a large number of amino acid pairs in *Hymenolepis diminuta* and in the intestinal mucosa of its rat host. Such data may furnish a more complete understanding of the number of qualitatively distinct amino acid transport loci in the parasite and allow an evaluation of differences between host and parasite.

MATERIALS AND METHODS

Organisms

Rat tapeworms, *Hymenolepis diminuta*, 10 days old ($\pm$ 4 hours) were used as standardized for transport studies by Read et al. (1963). Rat gut tissues were obtained from male albino Holtzman rats. The animals were maintained in the laboratory for about two weeks before use and weighed 160-190 g at the time of killing.

Methods with tapeworms

The inhibition of influx of each of 15 L-amino acids by each of the other 14 amino acids was determined using the methodology of Read et al. (1963). All incubations were of two-minute duration except for glutamic acid, with which one-minute incubations minimized metabolic efforts. Uninhibited influx was determined from incubations in 1 mM ^{14}C-amino acid in KRT (Krebs-Ringer tris-maleate buffered saline, Read et al., 1963); inhibited influx was determined from the uptake of 1 mM labeled substrate in the presence of 5 mM inhibitor in KRT. Influx in μmoles/gm ethanol-extracted dry weight/hr was determined by appropriate standards and calculations, from the radioactivity determined in the ethanol extract.

Percent efflux was determined by incubating the worms for one minute in 0.1 mM ^{14}C-labeled amino acid (in KRT), rinsing thrice in KRT, then placing the worms for two minutes in either KRT or 5 mM unlabeled amino acid in KRT, or directly into 70% ethanol. The amount of amino acid found in the sample placed directly into the ethanol was called the *initial influx*. The amount remaining in the samples after the subsequent two minutes in KRT was used to determine the passive efflux in KRT. The percentage of stimulated efflux was calculated by first subtracting the passive efflux of labeled amino acid leaked into KRT from the amount effluxed in the presence of 5 mM unlabeled amino acid in KRT, then dividing by the initial influx and multiplying by 100.

Other experiments with minor modifications in techniques are described in context.

Methods with Gut Tissues

The tissue accumulation method of Agar et al. (1954) was generally followed. Rats were killed by cervical dislocation, and a two-inch segment of the small intestine, the third to the fifth inch from the pylorus, was removed and immediately placed in cold Krebs-Ringer solution. The section was everted, rinsed, and cut into ring segments approximately $^1/_8''$ wide. The segments were rinsed twice in cold buffer, then placed in Krebs-Ringer's-bicarbonate buffer (pH 7.4) containing 5 mM glucose (called KRGB hereafter) for at least one minute at 37°C before incubation in test solutions.

Data were obtained in groups of sixteen samples from each rat, each $^1/_8''$ gut segment constituting a sample. From each rat four samples were used to determine uninhibited influx of labeled amino acid; the remaining twelve samples were divided into three groups of four to determine the inhibited influx in the presence of each of three different unlabeled amino acids. This procedure provided an internal control so that variation among rats did not significantly affect the data. The incubations were conducted in 50 ml beakers in a constant-temperature shaker bath at 37°C; 5% CO_2-95% air was gently bubbled through the solutions before and during incubation of the tissue. Uninhibited influx of labeled amino acid was determined from a one-minute incubation in 25 ml of 1 mM amino acid in KRGB. Inhibited influx was determined similarly, except that 5 mM inhibitor amino acid was present. After incubation the segment was rinsed thrice in buffered saline (room temperature), then placed in 70% ethanol to extract overnight. The amount of uptake was determined by removing a 0.5 ml aliquot of the ethanol extract and counting in a gas-flow Geiger counter. Conversion of radioactivity to μmoles was made with appropriate standards and calculations.

The uptake data for rat gut segments were based on total protein (serum-albumin equivalents remaining after ethanol extraction). Each gut segment was removed from the ethanol and digested in 5 ml of 1 N NaOH for one hour at 37°C. Total protein was determined by the method of Lowry et al. (1951).

To establish the validity of the experimental procedures with the rat gut tissues we made the following preliminary observations. We found that the uptake of alanine by each $\frac{1}{8}$" segment cut from a two-inch piece of gut was the same. We assumed that this was true for the other amino acids studied. A comparison by light and electron microscopy was made between gut tissue removed from the rat and immediately fixed in glutaraldehyde, and tissue which had undergone the experimental incubation before fixation. Tissue sections were prepared by the usual procedures for light and electron microscopy. Subsequent examination of the tissue revealed negligible effects attributable to the experimental procedures.

Statistical analyses

Correlation coefficients were calculated with the aid of an IBM 1620 computer, using a linear correlation coefficient program (IBM 6.0.038), as modified by John E. Simmons, Jr.

RESULTS AND CONCLUSIONS

Inhibition of Amino Acid Influx in *Hymenolepis*

We have determined the reciprocal inhibitory activity in the influx of amino acid pairs in all combinations of 15 amino acids into *Hymenolepis diminuta*. These data are summarized in figure 1. The vertical position of an amino acid in each column represents its activity as an inhibitor of the influx of the amino acid indicated at the base of the column. The columns in figure 1 were arranged to show relative increasing or decreasing inhibitory trends in the action of certain amino acids. Inspection of these data allows certain conclusions:

1. There is considerable overlapping in the affinity (inhibition) for various transport loci.

2. The most specific locus is the arginine-lysine system; histidine is the only other amino acid, of those tested, yielding a significant inhibition at this locus.

3. Methionine is a strong inhibitor at all transport loci other than the arginine-lysine locus.

4. There are obvious changes in the relative inhibiting activity of specific amino acids, looking from left to right in figure 1. These data show a progression from high inhibition by alanine, serine, threonine, and valine in the columns on the left to high inhibition by isoleucine, leucine, phenylalanine, and tyrosine on the right. The "crossing-over" trend of the inhibitions suggests the presence of more than one locus for the transport of monoamino-monocarboxylic acids, perhaps corresponding to the A and L systems described by Oxender and Christensen (1963) in Ehrlich ascites cells and by Read et al. (1963) in $H.$ $diminuta.$ Phenylalanine and histidine generally follow the trend of the L locus (leucine-preferring), except that phenylalanine is a better inhibitor of phenylalanine, tyrosine, and histidine influx than of leucine influx. From these observations, we concluded that there are at least two loci involved in the transport of neutral amino acids and a high probability of a third locus having a strong affinity for the aromatic amino acids.

A more precise way of examining the relative inhibitory activities of the amino acids studied involves the calculation of correlation coefficients. The coefficients for interactions of all pairs of 15 amino acids are shown in table 1. The inhibitory activities of phenylalanine and tyrosine on influx of other amino acids are more highly correlated ($r = 0.94 \pm 0.03$) than the inhibitions produced by leucine and phenylalanine ($r = 0.59 \pm 0.17$) or by leucine and tyrosine influx ($r = 0.54 \pm 0.18$). Leucine also shows a stronger association with alanine than does phenylalanine or tyrosine. Inhibitory effects of isoleucine are more highly correlated ($r > 0.90$) with effects of alanine, serine, threonine, methionine, valine, and leucine than with effects of phenylalanine or tyrosine ($r = 0.40$ and 0.35, respectively).

Also seen in table 1 is a high correlation ($r > 0.90$) between inhibitions effected by alanine and those by serine, threonine, methionine, isoleucine and, to a slightly lesser extent, valine ($r = 0.86$). These amino acids seem to react preferentially with a single locus. Isoleucine also seems to interact significantly with the locus involved in leucine influx.

In $H.$ $diminuta,$ proline has its highest correlation with valine. Inhibitions by glycine do not show high correlation with those of other amino acids tested, suggesting that there is a glycine-preferring locus; serine, threonine, and methionine clearly interact strongly with this locus (figure 1). Present and earlier data (Read et al., 1963) support the view that there is a monoamino-dicarboxylic acid-preferring locus.

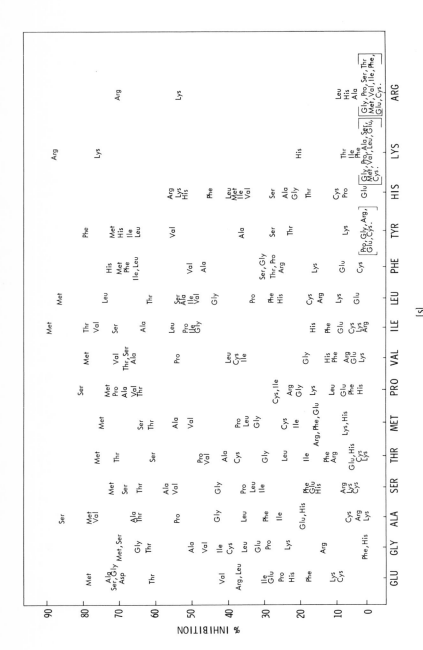

FIG. 1. COMPARISON OF INHIBITIONS by each of 15 amino acids on the uptake of amino acids by *H. diminuta*, substrate at 1 mM, inhibitor at 5 mM. The position of each inhibitor in the graph corresponds to the percentage of inhibition. Each inhibition was calculated from the mean of at least five samples (5 worms per sample). Uninhibited transport rates are listed in table 5.

TABLE 1

CORRELATION COEFFICIENTS OF THE PERCENTAGE OF INHIBITION BY 15 AMINO ACIDS ON VARIOUS AMINO ACIDS IN *H. DIMINUTA*

	Ala	Leu	Phe	Gly	Pro	Ile	Met	Ser	Thr	Val	Glu	Tyr	His	Arg	Lys
Glu	0.78	0.72	0.19	0.79	0.64	0.78	0.83	0.87	0.79	0.64	—	0.09	-0.11	-0.25	-0.30
Gly	0.68	0.65	-0.13	—	0.69	0.76	0.82	0.80	0.74	0.71	0.79	-0.12	-0.41	-0.38	-0.43
Ala	—	0.74	0.30	0.68	0.68	0.93	0.91	0.96	0.92	0.86	0.78	0.27	-0.21	-0.53	-0.56
Ser	0.96	0.83	0.30	0.80	0.83	0.97	0.95	—	0.95	0.88	0.87	0.27	-0.18	-0.48	-0.51
Thr	0.92	0.78	0.21	0.74	0.88	0.92	0.96	0.95	—	0.90	0.79	0.16	-0.20	-0.39	-0.42
Met	0.91	0.80	0.16	0.82	0.88	0.91	—	0.95	0.96	0.94	0.83	0.17	-0.26	-0.43	-0.49
Pro	0.68	0.55	-0.27	0.69	—	0.78	0.88	0.83	0.88	0.90	0.64	-0.55	-0.27	-0.27	-0.30
Val	0.86	0.78	0.23	0.71	0.90	0.90	0.94	0.88	0.90	—	0.64	0.25	-0.24	-0.49	-0.53
Ile	0.93	0.90	0.40	0.76	0.78	—	0.91	0.97	0.92	0.90	0.78	0.35	-0.16	-0.53	-0.57
Leu	0.74	—	0.59	0.65	0.55	0.90	0.80	0.83	0.78	0.78	0.72	0.54	0.61	-0.42	-0.48
Phe	0.30	0.59	—	-0.13	-0.27	0.40	0.16	0.30	0.21	0.23	0.19	0.94	0.54	-0.28	-0.26
Tyr	0.27	0.54	0.94	-0.12	-0.55	0.35	0.17	0.27	0.16	0.25	0.09	—	0.50	-0.34	-0.32
His	-0.21	0.61	0.54	-0.41	-0.27	-0.16	-0.26	-0.18	-0.20	-0.24	-0.11	0.50	—	0.59	0.62
Lys	-0.56	-0.48	-0.26	-0.43	-0.30	-0.57	-0.49	-0.51	-0.42	-0.53	-0.30	-0.32	0.62	0.98	—
Arg	-0.53	-0.42	-0.28	-0.38	-0.27	-0.53	-0.43	-0.48	-0.39	-0.49	-0.25	-0.34	0.59	—	0.98

There is evidence for a locus showing a strong preference for aromatic amino acids, which also exhibits a great affinity for leucine. Plotting the correlation coefficients for phenylalanine shows this clearly (figure 2). If the interactions of histidine with arginine and lysine are removed from consideration, the correlation of histidine with phenylalanine and tyrosine is much higher. Previous studies with a large series of aromatic amino acids furnish independent evidence for an aromatic amino acid-preferring locus (Read et al., 1963).

As an inhibitor, histidine (figure 1) is most effective against phenylalanine and tyrosine influx, but, unlike other neutral amino acids, histidine is also an effective inhibitor of lysine and arginine influx. The activity of other amino acids as inhibitors of histidine influx supports the conclusion of Woodward and Read (1969) that histidine is absorbed mainly through the aromatic and dibasic amino acid loci. In a separate experiment, we examined the interactions of phenylalanine, leucine, lysine, and arginine as inhibitors of histidine influx. The results (figure 3) show that when both a dibasic and a neutral amino acid are present as inhibitors, there is an additive inhibition of histidine influx, whereas the presence of two dibasic amino acids does not produce additional inhibition. This is further evidence that histidine has overlapping affinity for the otherwise highly specific aromatic and dibasic amino acid transport systems.

Since phenylalanine does not interact with the dibasic locus and arginine does not interact with the aromatic locus, we reasoned that it might be possible to demonstrate linear changes in the relative numbers of these two transport systems along the tapeworm strobila. Accordingly, single 14-day-old worms were incubated for two minutes in the presence of 1.0 mM ^{14}C-arginine and 1.0 mM ^{3}H-phenylalanine. After rinsing, each worm was stretched on a thin glass plate sitting in a tray of crushed dry ice. The worms were completely frozen in about a second. Each frozen worm was then cut into pieces about 2 mm in length and placed in 70% ethanol. The following day, radiocarbon and tritium levels were determined in each ethanol extract. The ratio of radiocarbon to tritium was constant along the entire length of the strobila in each of five worms examined by this procedure and we concluded that the relative numbers of these two systems appeared constant.

An additional observation may be made concerning the data in figure 1. The horizontal sequence of columns selected to show the maximum number of different trends in competitions resulted in an arrangement of the transported amino acids in a structural series. The dicarboxylic amino acid, glutamate, is on the left edge and the dibasic amino acids, lysine and arginine, are on the right edge. Between the extremities is a series from left to right that shows, respectively, an increase in length of the carbon chain, addition of a hydroxyl group to the chain, branching of the chain, and finally the ring structures of phenylalanine, tyrosine, and histidine. Proline may appear to be

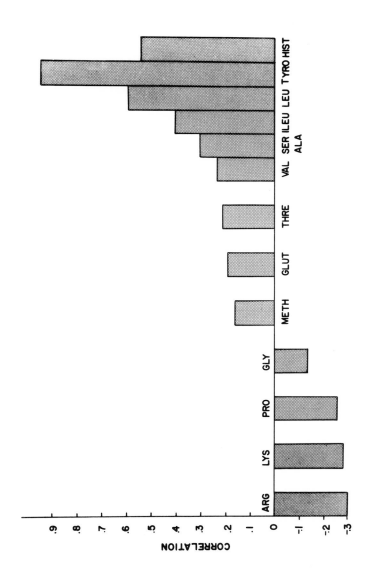

Fig. 2. Plot of correlation coefficients of inhibitions of each amino acid compared with inhibitions by phenylalanine in *H. diminuta.*

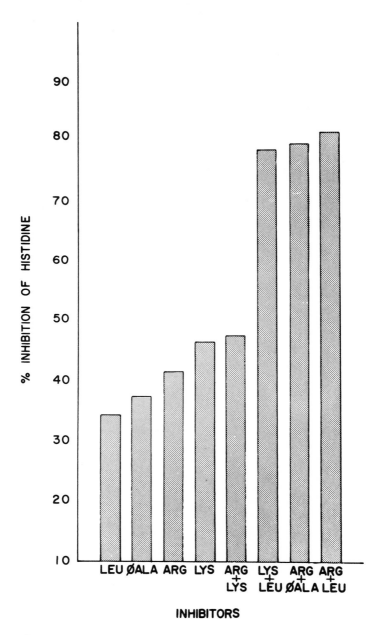

FIG. 3. COMPARISON OF THE EFFECTS of the presence of one (0.5 mM) or two (0.5 mM each) inhibitors on the uptake of histidine (0.1 mM) by *H. diminuta*.

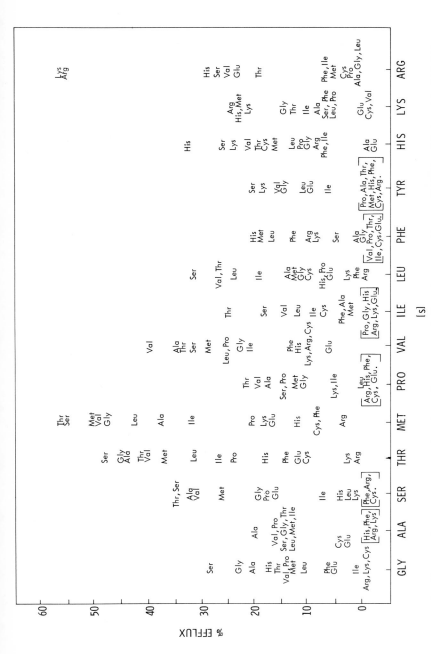

FIG. 4. COMPARISON OF EFFLUX of previously accumulated amino acid in the presence of single amino acids by *H. diminuta*. The position of each amino acid in the graph corresponds to the percentage of efflux (see methods for details). Each point was calculated from the mean of at least five samples.

an exception to this structural sequence. Its properties as an *imino* acid may account for its "fit" in the sequence, however, on the basis of the correlation coefficients of table 1, it seems logically placed between methionine and valine.

Efflux of Amino Acids from *Hymenolepis*

A number of investigations have shown that the presence of a solute in the surrounding medium may enhance the efflux of an analogous compound from the internal cellular pool. This has been shown to occur with amino acids in *H. diminuta* (Read et al., 1963; Hopkins and Callow, 1965; Kilejian, 1966; Woodward and Read, 1969; Arme and Read, 1969). We decided to examine systematically the relative activities of exogenous amino acids in enhancing outflow of previously absorbed amino acids. Such data might provide additional information on the specificity and overlapping affinities of transport loci and on the relative roles of loci in the efflux processes. The results of our experiments are shown in figure 4, and the correlation coefficients for the effects of individual external amino acids in enhancing efflux are given in table 2.

Generally, the same amino acid transport loci suggested by results from inhibition studies are supported by the data from efflux experiments. The interpretation of these data concerning the absolute value of efflux must be made with great caution. The absorbed amino acid was not, in any case, at a steady state level with regard to the internal pool. The relative activity of amino acids in promoting efflux should be valid for comparative purposes, however. Serine, threonine, alanine, and valine appear to be most effective in promoting efflux of other amino acids, suggesting that the locus shared by these amino acids is most active in efflux. Leucine and isoleucine are less effective than the above mentioned amino acids and phenylalanine is a weak effector of efflux. This suggests that the aromatic locus and the leucine-preferring locus are involved, to a relatively minor extent, in the efflux of neutral amino acids.

The efflux of glutamic acid was not examined because significant metabolism of this amino acid occurs in three minutes. The relative insolubility of tyrosine prevented examination of its activity as a stimulator of efflux at appropriate concentrations.

Inhibition of Amino Acid Influx in Rat Gut

The relative activities of various individual unlabeled amino acids as inhibitors of the influx of each of ten labeled amino acids into rat gut tissue are shown in figure 5. The correlation coefficients for the inhibitions produced among pairs of amino acids are presented in table 3. Although these data are limited, they clearly indicate the presence of more than one amino acid transport system and also suggest that there are overlapping affinities for amino acids by these systems. For example, while alanine is more effective

TABLE 2

CORRELATION COEFFICIENTS OF THE EFFECTS AS EXCHANGERS OF 14 L-AMINO ACIDS ON VARIOUS AMINO ACIDS IN *H. DIMINUTA*

	Phe	Leu	Ala	Ser	Gly	Pro	Thr	Val	Met	Ile	Arg	Lys	Tyr	His
Gly	-0.04	0.54	0.64	0.75	—	0.62	0.84	0.65	0.66	0.27	-0.32	-0.07	0.25	0.20
Ala	-0.45	0.69	—	0.78	0.64	0.80	0.87	0.91	0.78	0.43	-0.57	-0.35	0.15	-0.18
Ser	-0.38	0.65	0.78	—	0.75	0.90	0.83	0.82	0.81	0.52	-0.21	-0.19	0.20	0.05
Thr	-0.20	0.79	0.87	0.83	0.84	0.77	—	0.90	0.91	0.53	-0.51	-0.19	0.25	0.07
Met	-0.17	0.86	0.78	0.81	0.67	0.75	0.91	0.86	—	0.65	-0.30	-0.01	0.41	0.16
Pro	-0.44	0.58	0.80	0.90	0.62	—	0.77	0.87	0.75	0.51	-0.17	-0.10	0.15	0.11
Val	-0.26	0.78	0.91	0.82	0.65	0.87	0.90	—	0.86	0.62	-0.40	-0.24	0.17	0.06
Ile	-0.28	0.84	0.43	0.52	0.27	0.51	0.53	0.62	0.64	—	-0.11	-0.35	0.21	0.31
Leu	-0.23	—	0.69	0.65	0.54	0.57	0.79	0.78	0.86	0.84	-0.28	-0.34	0.42	0.30
Phe	—	-0.23	-0.45	-0.38	-0.04	-0.45	-0.20	-0.26	-0.16	-0.28	0.18	0.58	-0.20	0.33
Tyr	-0.20	0.42	0.15	0.21	0.25	0.15	0.25	0.17	0.41	0.21	0.29	-0.17	—	0.30
His	0.33	0.31	-0.18	0.05	0.20	0.11	0.07	0.06	0.16	0.31	0.38	0.28	0.30	—
Lys	0.59	-0.23	-0.35	-0.19	-0.07	-0.10	-0.19	-0.24	-0.01	-0.35	0.49	—	-0.17	0.28
Arg	0.18	-0.28	-0.57	-0.21	-0.32	-0.17	-0.51	-0.40	-0.31	-0.11	—	0.49	0.29	0.38

TABLE 3

Correlation Coefficients of the Inhibitory Action of 10 L-Amino Acids on Various Amino Acids in Rat Gut Tissue

	Glu	Gly	Ala	Ser	Met	Pro	Val	Leu	Phe	Lys
Glu	—	0.27	0.39	0.35	0.40	0.12	0.33	0.61	−0.24	0.38
Gly	0.27	—	0.11	−0.08	0.16	−0.25	0.13	0.11	−0.24	0.51
Ala	0.39	0.11	—	0.56	0.43	0.32	0.63	0.52	0.59	0.39
Ser	0.35	−0.08	0.56	—	0.70	0.30	0.67	0.36	0.62	0.08
Met	0.40	0.16	0.43	0.70	—	0.28	0.56	0.42	0.38	0.42
Pro	0.12	−0.25	0.32	0.30	0.28	—	0.58	0.39	0.07	0.09
Val	0.33	0.13	0.63	0.67	0.56	0.58	—	0.60	0.42	0.32
Leu	0.61	0.11	0.52	0.36	0.42	0.39	0.60	—	0.52	0.23
Phe	0.56	−0.24	0.59	0.62	0.38	0.07	0.42	0.52	—	0.29
Lys	0.38	0.51	0.39	0.08	0.42	0.09	0.32	0.23	0.29	—

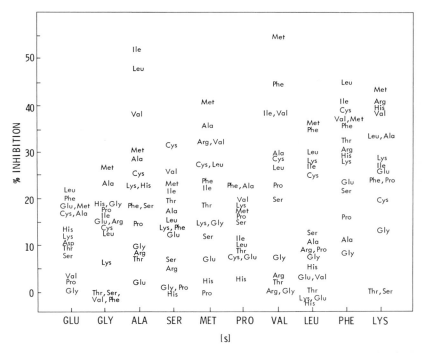

FIG. 5. COMPARISON OF INHIBITIONS by each of 15 amino acids on the uptake of various amino acids in segments of rat gut tissue. The position of each inhibitor in the graph corresponds to the percentage of inhibition. Substrate concentration at 1 mM, inhibitor at 5 mM. Each inhibition was calculated from the mean of at least four samples. Uninhibited transport rates are listed in table 5.

than leucine as an inhibitor of glycine, methionine, and proline influx, the converse is true in the relative effects of leucine and alanine on the influx of leucine, alanine, and phenylalanine. Likewise, valine and alanine have very similar effects on proline influx, but these two amino acids differ sharply in their effects on glycine influx.

We may recall that lysine clearly has overlapping affinity with systems involved in the transport of neutral amino acids, as reported previously by Arme and Read (1969). Also, glycine and lysine show a higher correlation coefficient with one another for the inhibition of the influx of other amino acids than either of them shows with any other amino acid tested. Hampton (1970) found lysine to interact strongly with the neutral amino acid transport systems in *Trypanosoma cruzi*, a protozoan parasite of mammals. These systems differ sharply from *Hymenolepis diminuta*'s. The marked inhibition of lysine influx produced by arginine and histidine suggests that there is a locus showing strong affinity for basic amino acids.

Proline shows a low correlation coefficient with most amino acids tested, with greatest relation to the branched chain amino acids, valine and leucine. Note in figure 5 that leucine, isoleucine, valine, and cysteine follow the same general pattern as inhibitors of the influx of other amino acids.

DISCUSSION

Examination of the relative effects of various individual amino acids on the influx and efflux of other amino acids indicates that *Hymenolepis diminuta* has several amino acid transport loci, differing qualitatively but with considerable overlapping affinity for various amino acids. From our data and previous studies, we can postulate six amino acid transport systems for *H. diminuta*, as summarized in table 4.

In rat gut, there appear to be an alanine-preferring system, a leucine-preferring system, a glycine-preferring system, and a basic amino acid-preferring system. More extensive study is required to identify other systems that may be present. As with *Hymenolepis*, there is considerable overlap of affinity for the amino acids examined. The proline-preferring system studied by Hagihira et al. (1961) is probably identical with what we have referred to as the leucine-preferring system. As shown by Hagihira et al., this system exhibits affinity for branched chain compounds that may be regarded, in the broad sense, as analogues of leucine, valine, or isoleucine.

It is also interesting to compare the amino acid transport systems of the symbiotic tapeworm *Hymenolepis diminuta* with those of the host's intestinal mucosa. First, we must point out that our studies of the rat intestinal mucosa were made with unparasitized host tissue. Although we recognize that the presence of the tapeworm might modify the characteristics of transport systems of the mucosa, it seems probable that they would only be changed in the quantitative sense. With that assumption, there are clearly very significant differences between the parasite and the intestinal mucosa.

The most striking differences between the inhibitions in the rat and worm are the effects of leucine and isoleucine on alanine uptake. It may be seen from the data obtained with the rat gut segments (figure 5) that leucine and isoleucine are effective inhibitors not only of leucine but also of alanine (cf. figure 1). In the rat gut, alanine is not as good an inhibitor of leucine or phenylalanine as it is in the worm. In the rat gut, valine is a relatively good inhibitor of alanine, but not of leucine, yet valine appears as a good inhibitor of phenylalanine. Valine's effect as an inhibitor of the various amino acids does not follow the same pattern in the worm. These results demonstrate an obvious difference in the nature of the A and L loci in host and parasite.

The screening of interactions in amino acid transport by examining single inhibitor-to-substrate ratios is useful, but this approach has limitations. For example, it furnishes no evaluation of whether there may be multiple binding

TABLE 4

Six Amino Acid Transport Systems for *H. Diminuta*

LOCI	DICARBOXYLIC	GLYCINE	SERINE	LEUCINE	PHENYLALANINE	DIBASIC
MAIN AMINO ACIDS USING THE LOCUS	aspartic glutamic	glycine	serine alanine threonine	leucine isoleucine	phenylalanine tyrosine histidine	arginine lysine
	methionine	methionine	methionine valine proline	methionine	methionine	
OVERLAPPING AMINO ACIDS	serine alanine glycine	serine threonine alanine	glycine	glycine serine threonine alanine valine	leucine isoleucine	histidine

sites on one or more systems. Further, it does not detect other kinds of inter-action, such as stimulation, which may occur at other inhibitor: substrate ratios. Such interactions are known to occur in *Hymenolepis* in the absorption of purines and pyrimidines (MacInnis et al., 1965; MacInnis and Ridley, 1969; Pappas et al., 1973) and of long chain fatty acids (Chappell et al., 1969). Ruff and Read (1974) have recently demonstrated such interactions in the absorption of amino acids by the hemoflagellate *Trypanosoma equiperdum.*

The screening method used in our study does not distinguish non-productive binding, i.e., binding without subsequent transport, which is known to occur with purines and pyrimidines in *H. diminuta* (Pappas et al., 1973), with amino sugars in *Trypanosoma gambiense* (Southworth and Read, 1972), and with sugars in *Schistosoma mansoni* (Uglem, personal communication). If a compound binds but is not transported through a particular locus, it may still exhibit competitive inhibition. This may account for the discrepancies between K_t (Michaelis constant) and K_i (inhibition constant) observed in amino acid transport in *H: diminuta* by Read et al. (1963). K_i values could be used to predict the rate of absorption of a single component from a mixture of amino acids, whereas K_t values did not yield satisfactory predictions. This is exactly what would be expected if some amino acids undergo non-productive binding on certain transport loci. Thus the free pools of amino acid in the gut may influence the nature of competitive inhibitions, non-productive binding yielding inhibition and other factors not yet delin-eated. Of the latter, we can predict that energy sources such as glucose will also be of importance. CO_2, pH, NH_4, Na, K, and Ca will also impinge on this competition. What effect do these items have in regulating protein synthesis? Are they coupled to the regulation of RNA synthesis? Questions such as these must be answered before a satisfactory model of host-parasite interaction can be proposed. A beginning on this complex problem may be attempted, however, by summarizing some of the data we have accumulated on amino acids. In table 5 we have listed, in order of rank, various properties and effects of amino acids in the tapeworm and in the rat's gut. Presently, they suggest at least one obvious conclusion: The tapeworm has evolved remarkably well-adapted systems that favor its growth in competition with the host. A look at table 5 reveals, for example, that the rank orders of the amino acids as inhibitors in the worm and rat are not correlated. One would expect that many of the parameters examined would be correlated with the size of the free pools. Yet no simple correlations of pool size with exchange rate, initial uptake, or protein synthesis are evident. These data should simplify future studies by delineating aspects most likely to provide decisive information on host-parasite competition. Perhaps they will also reveal patterns of value to future investigators. One of the most important, as yet unanswered, questions concerning amino acid flux is whether or not the tapeworm modifies its environment in ways that benefit its own welfare.

TABLE 5
COMPARISON OF RANK ORDER OF VARIOUS PROPERTIES AND EFFECTS OF AMINO ACIDS IN THE TAPEWORM AND RAT GUT TISSUE

Uptake rates by *H. diminuta* from 0.1 mM substrate are from one-minute incubations used to calculate efflux rates shown in table 4. Protein synthesis rates were determined from aliquots of TCA precipitates of homogenized worm tissue.

	H. diminuta Uptake Rate S = 0.1 mM μmoles / gm / hr	*H. diminuta* Uptake Rate S = 1.0 mM μmoles / gm / hr	Rat Gut Uptake Rate S = 1.0 mM μmoles / gm / hr	*H. diminuta** Free Pool	Rat Gut* Lumen	*H. diminuta* $\bar{x}$% Inhibition of Other Amino Acids	Rat Gut $\bar{x}$% Inhibition of Other Amino Acids	*H. diminuta*** % of Total Uptake From 2 hr Incubation Incorporated Into Protein
1	met 57	thr 167	met 169	ala	glu	met 64	met 32	ile 74
2	thr 57	ala 164	val 147	glu	gly	thr 49	ile 28	phe 55
3	val 56	val 140	phe 146	gly	ala	val 48	leu 26	thr 55
4	ala 43	met 139	leu 144	pro	lys	ser 46	phe 23	val 52
5	ser 39	ile 130	ser 134	ser	asp	ala 41	cys 23	arg 50
6	leu 32	pro 127	pro 121	leu	ser	leu 36	val 23	lys 44
7	pro 31	ser 122	gly 120	lys	leu	ile 32	ala 23	leu 38
8	gly 27	gly 117	ala 92	val	arg	gly 32	lys 19	tyr 37
9	arg 23	leu 92	glu 65	asp	pro	pro 32	arg 15	ser 36
10	his 19	phe 86	lys 57	his	val	arg 25	his 14	his 32
11	ile 19	his 51		ile	thr	his 24	pro 12	asp 24
12	lys 18	tyr 37		met	ile	lys 21	ser 11	gly 16
13	phe 18	lys 34		phe	phe	phe 21	glu 11	pro 15
14	tyr 14	arg 26		thr	his	glu 10	thr 9	ala 8
15	glu	glu 18		arg	met		gly 8	glu 6

*From J. E. Simmons **MacInnis, Graff, Fisher and Read, unpublished $\bar{x}$% Average percentage

ACKNOWLEDGMENTS

This work was supported in part by grants from the National Institutes of Health (AI 13830, AI 01384, 5 T01 AI 00106 to Clark P. Read; AI 00070 and NSF 8753 to A. J. MacInnis).

We wish to express our deep gratitude to Mr. William Kitzman for his exceptional assistance with these experiments. His skills, comradeship, and good humor provided a unifying force over nearly two decades in Clark Read's laboratory. To Evelyn Hake we extend similar thanks for her selfless devotion to science, assistance with these experiments, and innumerable contributions to the parasitology group at Rice beginning in the Chandler era.

REFERENCES CITED

Agar, W. T., F. J. R. Hird, and G. S. Sidhu
  1954 The uptake of amino acids by the intestine. Biochimica et Biophysica Acta 14:80-84.

Arme, C. and C. P. Read
  1969 Fluxes of amino acids between the rat and a cestode symbiote. Comparative Biochemistry and Physiology 29:1135-1147.

Chappell, L. H., C. Arme, and C. P. Read
  1969 Studies on membrane transport V. Transport of long chain fatty acids in *Hymenolepis diminuta* (cestoda). Biological Bulletin 136:313-326.

Hagihira, H., E. C. C. Lin, A. H. Samiy, and T. H. Wilson
  1961 Active transport of lysine, ornithine, arginine and cystine by the intestine. Biochemical and Biophysical Research Communications 4:478-481.

Hampton, J. R.
  1970 Lysine transport in the culture form of *Trypanosoma cruzi*: Kinetics and inhibition of uptake by structural analogues. International Journal of Biochemistry 1:706-714.

Hopkins, C. A. and L. L. Callow
  1965 Methionine flux between a tapeworm (*Hymenolepis diminuta*) and its environment. Parasitology 55:653-666.

Kilejian, A.
  1966 Formation of the L-proline pool of the cestode, *Hymenolepis diminuta*. Experimental Parasitology 19:358-365.

Lowry, O. H., N. J. Rosebrough, A. L. Farr, and R. J. Randall
  1951 Protein measurement with the Folin phenol reagent. Journal of Biological Chemistry 193:265-275.

MacInnis, A. J., F. M. Fisher, Jr., and C. P. Read
1965 Membrane transport of purines and pyrimidines in a cestode. Journal of Parasitology 51:260-267.

MacInnis, A. J. and R. K. Ridley
1969 The molecular configuration of pyrimidines that causes allosteric activation of uracil transport in *Hymenolepis diminuta*. Journal of Parasitology 55:1134-1140.

Oxender, D. L. and H. M. Christensen
1963 Evidence for two types of mediation of neutral amino acid transport in Ehrlich cells. Nature 197:765-767.

Pappas, P. W., G. L. Uglem, and C. P. Read
1973 The influx of purines and pyrimidines across the brush border of *Hymenolepis diminuta*. Parasitology 66:525-538.

Read, C. P., A. H. Rothman, and J. E. Simmons
1963 Studies on membrane transport with special reference to host-parasite integration. Annals of the New York Academy of Science 113:154-205.

Rothman, A. H. and F. M. Fisher, Jr.
1964 Permeation of amino acids in *Moniliformis* and *Macracanthorhynchus* (Acanthocephala). Journal of Parasitology 50:410-414.

Ruff, M. D. and C. P. Read
1974 Interactions of amino acids in membrane permeation in *Trypanosoma equiperdum*. In preparation.

Senturia, J. B.
1964 Studies on the absorption of methionine by the cestode, *Hymenolepis citelli*. Comparative Biochemistry and Physiology 12:259-272.

Southworth, G. C. and C. P. Read
1972 Absorption of some amino acids by the haemoflagellate, *Trypanosoma gambiense*. Comparative Biochemistry and Physiology 41A:905-911.

Winters, C. G. and H. N. Christensen
1964 Migration of amino acids across the membrane of the human erythrocyte. The Journal of Biological Chemistry 239:872-878.

Woodward, C. K. and C. P. Read
1969 Studies on membrane transport VII. Transport of histidine through two distinct systems in the tapeworm *Hymenolepis diminuta*. Comparative Biochemistry and Physiology **30**:1161-1177.

EFFECTS OF CARBON DIOXIDE
ON GLUCOSE INCORPORATION IN FLATWORMS

by James S. McDaniel, Austin J. MacInnis, and Clark P. Read

ABSTRACT

The rates of glucose-^{14}C incorporation into polysaccharide under CO_2 and CO_2-limited atmospheres were determined in flatworms of three Orders of Turbellaria, three of Cestoda, and one of Trematoda. Carbon dioxide inhibited glucose incorporation in *Neochildia fusca* (Acoela), a free-living species, and in the ectosymbiotes *Bdelloura candida* (Tricladida) and *Stylochus zebra* (Polycladida). The presence of carbon dioxide increased glucose incorporation in the parasitic tapeworms *Lacistorhynchus tenuis* (Trypanorhyncha), *Calliobothrium verticillatum* and *Orygmatobothrium dohrnii* (Tetraphyllidea), and *Tetrabothrius erostris* (Cyclophyllidea) and the trematode *Cryptocotyle lingua* (Digenea). The results suggest that the rates of glycogen synthesis in some flatworms vary with the level of available carbon dioxide in the environment.

INTRODUCTION

It has been reported that carbon dioxide has marked effects on carbohydrate metabolism in parasitic flatworms (Fairbairn et al., 1961; Kilejian, 1963; Read, 1967), and carbon dioxide fixation has been reported in both free-living and parasitic flatworms (Prescott and Campbell, 1965; Flickinger, 1959; Hammen and Lum, 1962). Kilejian (1963) reported that carbon dioxide was without significant effect on glucose uptake or net glycogenesis in the acanthocephalan *Moniliformis dubius*, although lack of carbon dioxide sharply depresses these functions in some tapeworms (Fairbairn et al., 1961; Read, 1967), and slightly depresses them in others (Von Brand and Stites, 1970). It thus seemed desirable to ascertain the effects of carbon dioxide on carbohydrate metabolism in a representative group of flatworms. Since some flatworms do not synthesize measurable glycogen in short-term experiments,

---

James McDaniel is Professor of Biology at East Carolina University. Austin MacInnis is Professor of Biology at the University of California, Los Angeles. Clark P. Read is deceased.

the incorporation of uniformly labeled glucose-^{14}C into worm glycogen was examined.

MATERIALS AND METHODS

Free-living and symbiotic flatworms were collected in the vicinity of Woods Hole, Massachusetts. The acoel turbellarian *Neochildia fusca* was collected from mud samples. *Bdelloura candida*, a triclad symbiote of the horseshoe crab *Limulus polyphemus*, was collected from the gill books of naturally infected hosts. *Stylochus zebra*, a polyclad ectosymbiote of the hermit crab (*Pagurus pollicaris*), was collected from crab-inhabited *Busycon* shells. The trypanorhynch tapeworm *Lacistorhynchus tenuis* and the tetraphyllidean tapeworm *Calliobothrium verticillatum* were collected from naturally infected dogfish (*Mustelus canis*). The tetraphyllidean *Orygmatobothrium dohrnii* was collected from the sand shark (*Carcharias taurus*). The cyclophyllidean tapeworm *Tetrabothrius erostrus* and the digenean trematode *Cryptocotyle lingua* were collected from the herring gull (*Larus argentatus*).

The flatworms were incubated in the following solutions to which sufficient sodium bicarbonate was added to maintain a pH of 7.4 when the animals were gassed with carbon dioxide: *Neochildia*, *Bdelloura*, and *Stylochus* in artificial seawater (Cavanaugh, 1964) containing 6 mM tris-maleate buffer (pH 7.4); *Lacistorhynchus*, *Calliobothrium*, and *Orygmatobothrium* in elasmobranch saline containing 25 mM tris-maleate buffer at pH 7.4 (Read et al., 1960); and *Tetrabothrius* and *Cryptocotyle* in Ringer's solution containing 25 mM tris-maleate buffer at pH 7.4. For carbon dioxide-limited incubations, bicarbonate was omitted and 30% KOH was added to a center well in each flask. All incubation media consisted of 20 ml containing 40 micromoles of glucose uniformly labeled with 1 microcurie of ^{14}C. Incubations were carried out at 20°C, except for those involving the avian parasites, *Tetrabothrius* and *Cryptocotyle*, which were performed at 38°C.

At the end of incubations, worms were quickly rinsed three times in 150 ml of the appropriate salt solution, blotted on hard filter paper, placed in a measured volume of 80% ethanol, and extracted for at least 24 hours. The ethanol-extracted worms were dried at 95°C for 12 hours, weighed, and dissolved in warm 1% NaOH. Aliquots were used for estimation of protein by the method of Lowry et al. (1951). To the remaining material KOH was added, to a concentration of 20%. After warming to 75°C, ethanol was added, to a final concentration of 50%. The precipitated alkali-stable polysaccharide was redissolved in water and an aliquot removed for carbohydrate estimation. The remaining polysaccharide was reprecipitated in 50% ethanol five times and aliquots removed for counting of radioactivity and for carbohydrate estimation. Carbohydrate was determined by the method of Dubois et al. (1956).

RESULTS AND DISCUSSION

The incorporation of glucose carbon in the species of flatworms studied is presented in table 1, along with data on the total ethanol-precipitable carbohydrate found in these forms. Carbon dioxide had effects on glucose incorporation in all cases. In all three turbellarians, more glucose was incorporated in the absence than in the presence of carbon dioxide, while the four cestodes and one trematode showed higher rates of glucose incorporation in the presence of carbon dioxide. Among these species, carbon dioxide fixation is known to occur in *Bdelloura* and *Stylochus* (Hammen and Lum, 1962). McDaniel and Dixon (1967) showed carbon dioxide to have an inhibitory effect on glucose incorporation in redia of *C. lingua*, whereas the opposite effect occurs in the adult.

The present data suggest that the significance of carbon dioxide fixation in the metabolism of these forms is not identical. The disparity may be related to significant differences in the terminal electron transport mechanisms of the parasites. The cestodes and the trematode examined in these experiments live in the digestive tract of vertebrates, probably under very low oxygen and high carbon dioxide tensions. The turbellarians, on the other hand, may live under relatively high oxygen tensions with modest carbon dioxide tensions. The free-living acoel, *N. fusca*, is found on the surface of mud bottoms. When the oxygen tension begins to drop, however, as in a collected bucket of mud, they quickly move away from the mud surface. If the bucket of mud is gently aerated, they remain on the mud surface.

TABLE 1
EFFECT OF CARBON DIOXIDE ON GLUCOSE-^{14}C
INCORPORATION IN FLATWORMS

SPECIES	ALKALI-STABLE POLYSACCHARIDE[a]	INCORPORATION[b]		
		No CO$_2$	Plus CO$_2$	Percentage of change
Neochildia fusca	791.1 (3)[c]	4.4	1.8	− 59
Bdelloura candida	234.0 (2)	8.5	6.7	− 21
Stylochus zebra	270.2 (4)	6.2	4.5	− 27
Lacistorhynchus tenuis	930.2 (1)	173.7	219.7	+ 27
Calliobothrium verticillatum	1069.6 (4)	245.2	323.4	+ 32
Orygmatobothrium dohrnii	514.3 (3)	17.8	43.5	+144
Tetrabothrius erostrus	5173.4 (3)	42.7	50.8	+ 19
Cryptocotyle lingua	762.6 (1)	21.6	36.4	+ 69

a. Expressed as nMoles glucose/mg. protein
b. Radioactivity as nMoles glucose/μMole glycogen/2 hrs.
c. Number of replicates in parentheses.

The findings in the present experiments suggest the difficulty of interpreting a general phenomenon, such as carbon dioxide fixation, with regard to its significance in the overall metabolism or ecology of a specific organism, or organisms. In addition they reveal possible regulatory phenomena worthy of future, more detailed investigation.

ACKNOWLEDGMENT

This work was supported in part by a grant from the National Institutes of Health, U. S. Public Health Service (AI-01384).

REFERENCES CITED

Cavanaugh, G. M., ed.
   1964 Formulae and Methods V. Marine Biological Laboratories, Woods Hole, Mass.

Dubois, M., K. A. Gilles, J. K. Hamilton, P. A. Rebers, and F. Smith
   1956 Colorimetric method for determination of sugars and related substances. Analytical Chemistry **28**:350-356.

Fairbairn, D., G. Wertheim, R. P. Harpur, and E. L. Schiller
   1961 Biochemistry of normal and irradiated strains of *Hymenolepis diminuta*. Experimental Parasitology **11**:248-263.

Flickinger, R. A.
   1959 A gradient of protein synthesis in planaria and reversal of axial polarity of regenerates. Growth **23**:251-271.

Hammen, C. S. and S. C. Lum
   1962 Carbon dioxide fixation in marine invertebrates. III. The main pathway in flatworms. Journal of Biological Chemistry **237**:2419-2422.

Kilejian, A. [printed as *Kilejan*]
   1963 The effect of carbon dioxide on glycogenesis in *Moniliformis dubius* (Acanthocephala). Journal of Parasitology **49**:862-863.

Lowry, O. H., N. J. Rosebrough, A. L. Farr, and R. J. Randall
   1951 Protein measurement with the Folin phenol reagent. Journal of Biological Chemistry **193**:265-275.

McDaniel, J. S. and K. E. Dixon
   1967 Utilization of exogenous glucose by the rediae of *Parorchis*

acanthus (Digenea: Philophthalmidae) and *Cryptocotyle lingua* (Digenea: Heterophyidae). Biological Bulletin **133**:591-599.

Prescott, L. M. and J. W. Campbell
  1965 Phosphoenolpyruvate carboxylase activity and glycogenesis in the flatworm, *Hymenolepis diminuta*. Comparative Biochemistry and Physiology **14**:491-511.

Read, C. P.
  1967 Carbohydrate metabolism in *Hymenolepis* (Cestoda). Journal of Parasitology **53**:1023-1029.

Read, C. P., J. E. Simmons, J. W. Campbell, and A. H. Rothman
  1960 Permeation and membrane transport in animal parasitism: Studies on a tapeworm-elasmobranch symbiosis. Biological Bulletin **119**:120-133.

Von Brand, T. and E. Stites
  1970 Aerobic and Anaerobic Metabolism of Larval and Adult *Taenia taeniaeformis*. VI. Glycogen synthesis from single substrates and substrate mixtures. Experimental Parasitology **27**:444-453.

STUDIES ON BIOCHEMICAL PATHOLOGY IN TRICHINOSIS. II. CHANGES IN LIVER AND MUSCLE GLYCOGEN AND SOME BLOOD CHEMICAL PARAMETERS IN MICE

by George L. Stewart

ABSTRACT

Glycogen content of mouse diaphragm muscle infected with *Trichinella spiralis* increased above that of uninfected mouse muscle early during a 34-day period following infection, and subsequently decreased to levels found in control muscle after day 24 post-infection (PI). There was no significant difference between the ratio of infected and uninfected mouse liver wet weight : dry weight, total liver protein, ratio of liver wet or dry weight : mouse body weight, or liver glycogen. Total body weight of trichinella-infected mice was less than that of uninfected mice after day 10 PI. The level of serum lactate in trichinous mice was above that of uninfected mice between days 6 and 12 PI. Serum glucose levels in infected animals were less than those in uninfected control mice between days 4 and 14 PI, and there was no significant difference in serum insulin levels between infected and uninfected mice over a 25-day period of study. There was a significant decrease in the concentration of blood urea nitrogen of trichinella-infected mice below that of control mice between days 4 and 24 PI.

INTRODUCTION

On the light and electron microscopic level, a number of workers have observed an increase in the amount of glycogen present in skeletal muscle infected with *Trichinella spiralis* (Zarzycki, 1956; Fasske and Themann, 1961; Beckett and Boothroyd, 1962; Karpiak et al., 1963). These reported increases in muscle glycogen occurred before day 20 post-infection and were followed by a decrease in muscle glycogen to levels similar to, or below, those of uninfected muscle. In the present study these alterations in muscle glycogen

---

George Stewart is Lecturer in Biology at Rice University.

211

were demonstrated by chemical means and a number of associated physical and chemical parameters in blood and liver from mice infected with *T. spiralis* were observed.

MATERIALS AND METHODS

The source of trichinella, method of excystment of muscle larvae and procedure for infection of mice were those used in previous studies (Stewart and Read, 1972a). Male or female 6-week-old Swiss white mice (Texas Inbred Mice Co., Houston, Texas) were used in all experiments. Mice were infected with 1000 larvae.

The anthelminthic methyridine was given to all infected and uninfected mice on day 11 PI, unless otherwise stated in context (Stewart and Read, 1972a).

For extraction of glycogen from diaphragm muscle and liver, tissue samples were removed at 2-minute intervals, rapidly weighed, and immediately immersed in 2 ml of 30% KOH at 100°C. All samples were removed from the boiling water bath after 15 minutes and cooled to room temperature. Glycogen was extracted and determined by the method of Montgomery (1957).

In determination of glycogen in trichinella larvae, worms were excysted on the days stated in context, washed 3 times in 0.85% NaCl (saline), separated into three equal samples and counted. Following centrifugation and removal of the supernatant (saline), 1 ml of 30% KOH at 100°C was pipetted onto the larval pellet. Each suspension of larvae was treated as above for extraction and determination of glycogen.

For removal of worms from muscle prior to encystment (before day 24 PI), the total body musculature of infected mice was run through a meat grinder, placed in saline solution in stoppered flasks with glass beads and shaken for 15 minutes at medium speed. The resulting suspension was strained through several thicknesses of cheesecloth and collected larvae were washed three times in saline.

The number of larvae/mg dry weight of diaphragm muscle was determined on 20 mice from the group of animals used in the study on muscle glycogen by methods previously outlined (Stewart and Read, 1972a).

For determination of serum lactate and serum glucose, 1 ml of blood was drawn, by direct heart puncture, from each experimental animal, and immediately placed in 2 ml of 0.3N BaOH. Two ml of 5% $ZnSO_4$ were subsequently added and the resulting precipitate was centrifuged at 2000 rpm for 15 minutes. One ml of the supernatant was used for serum lactate determination by the method of Barker-Summerson (1941), and 0.5 ml of the supernatant was used in the determination of serum glucose by the "Glucostat Special" method (Worthington Biochemical Corp., Freehold, N.J., 07728).

In determination of mouse liver protein, a sample of liver was removed, rinsed briefly in cold saline, blotted dry on filter paper, weighed, and homogenized at 0°C in 2 ml of glass-distilled water. Proteins were precipitated at 5°C by addition of an equal volume of 10% trichloroacetic acid. After centrifugation the protein pellet was dissolved in 1N NaOH and total protein was determined by the method of Lowry et al. (1951).

In studies on muscle and liver glycogen, blood glucose, lactate, urea nitrogen, and insulin, all animals were starved for 12 hours before killing. All animals in experiments dealing with chemical determinations were killed between 10:00 A.M. and 11:00 A.M. on the days indicated in context, and death was induced by intraperitoneal injection of 0.5 ml of 0.75% chlorobutanol (Eastman Kodak Co., Rochester, N.Y.).

Adult trichinella were collected by the 0.05% NaOH method of Larsh et al. (1952).

Blood urea nitrogen was determined by a modification of the Gentzkow-Masen (1942) method (Sigma Chemical Co., P. O. Box 14508, St. Louis, Mo., 63178).

Blood insulin was determined using an "Insulin Immunoassay Kit" (Amersham/Searle Corp., 2636 S. Clearbrook Dr., Arlington Heights, Illinois, 60005).

For determination of liver dry weights, whole livers were removed, placed in preweighed aluminum foil cups and put in a 90°C oven for 72 hours.

All chemicals were reagent grade. Student's "t" test was employed to evaluate the significance of differences.

RESULTS

Experiment I

Total glycogen was determined every third day between days 4 and 16 post-infection (PI) and on days 18, 20, 24, 28, 32, and 34 PI in diaphragm muscles from mice infected with *Trichinella spiralis* and uninfected mice. Total μg glycogen of diaphragm muscles from trichinella-infected mice was an average of 55.26% (SE ± 3.10) above that of uninfected after day 13 PI (figure 1). Total glycogen was determined for trichinella larvae (table 1) between days 16 and 34 PI and subtracted from total glycogen of infected diaphragm muscle (figure 1, line A). Results are presented in figure 1, line B. Total glycogen from infected mouse diaphragm muscles adjusted for the larval component is significantly different from that of uninfected mice between days 16 and 24 PI (an average increase of 38.95%; S.E. ± 4.90).

Experiment II

Mg% glucose in blood from infected and normal mice was determined every other day between days 2 and 24 PI (samples from day 22 PI were lost).

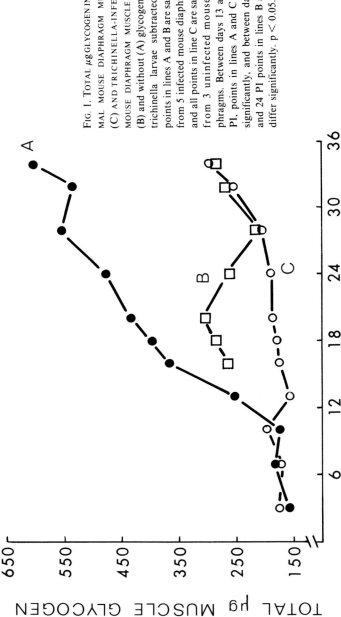

FIG. 1. TOTAL μg GLYCOGEN IN NOR-MAL MOUSE DIAPHRAGM MUSCLE (C) AND TRICHINELLA-INFECTED MOUSE DIAPHRAGM MUSCLE with (B) and without (A) glycogen from trichinella larvae subtracted. All points in lines A and B are samples from 5 infected mouse diaphragms and all points in line C are samples from 3 uninfected mouse dia-phragms. Between days 13 and 34 PI, points in lines A and C differ significantly, and between days 16 and 24 PI points in lines B and C differ significantly. p < 0.05.

TABLE 1

TOTAL μg GLYCOGEN OF TRICHINELLA LARVAE
ON DAYS 16, 18, 20, 24, 28, 32, AND 34 PI
Standard errors are listed in adjacent column. Determinations were made
on 3 samples of larvae on each day indicated.

Day PI	Average μg glycogen per larva	S.E.
16	0.0251	± 0.0017
18	0.0283	± 0.0017
20	0.0291	± 0.0023
24	0.0398	± 0.0014
28	0.0541	± 0.0023
32	0.0590	± 0.0013
34	0.0595	± 0.0017

TABLE 2

TOTAL NUMBER OF ADULT TRICHINELLA RECOVERED FROM THE
GUTS OF MICE NOT GIVEN METHYRIDINE
Data presented as percentage of total number of worms in inoculum (1000).
Standard errors are in an adjacent column. Determinations were made on
3 mice on each day in the experiment.

Day PI	Number of worms recovered ÷ 1000(%)	S.E.
8	87.9%	± 2.69
12	55.4%	± 4.32
14	46.4%	± 4.08
16	31.2%	± 5.26
18	17.1%	± 3.09
20	9.4%	± 0.91
24	0.3%	± 0.058

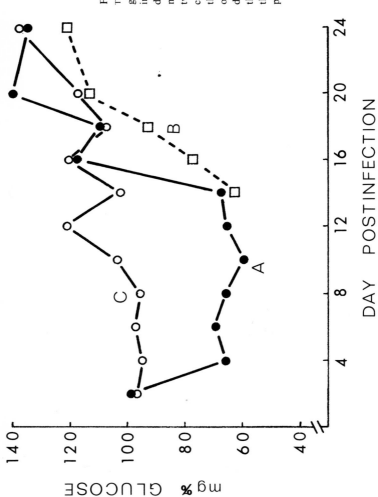

FIG. 2. Mg% BLOOD GLUCOSE FROM TRICHINELLA-INFECTED MICE not given methyridine (B), trichinella-infected mice given methyridine on day 11 PI (A), and uninfected animals (C). All points in line A between days 4 and 14 PI are significantly different from points from the same days in line C, and points on days 14, 16, and 18 PI in line B differ significantly from points on these days in line C. All points are the average of samples from 3 mice. p < 0.05.

Mg% blood glucose in infected mice given methyridine (figure 2, line A) on day 11 PI decreased between days 4 and 14 PI an average of 38.6% (S.E. ± 5.27) below that of uninfected animals (figure 2, line C). After day 14 PI, Mg% blood glucose in infected mice not given methyridine (figure 2, line B) remained below that of uninfected mice (figure 2, line C) through day 18 PI. No adult trichinella were found after day 12 PI in the guts of infected mice given methyridine. Percentage of total worms in the infecting dose recovered from mice not given methyridine is shown in table 2.

Mg% lactate in blood from infected and uninfected mice was determined in blood samples from the same mice used in blood glucose determinations. Mg% blood lactate in infected mice given methyridine on day 11 PI (figure 3, line A) rose between days 6 and 12 PI an average of 37% (S.E. ± 1.73) above that of control animals (figure 3, line C). After day 12 PI, blood lactate in infected animals not given methyridine remained above that of uninfected mice between days 14 and 18 PI (average increase = 20.2%; S.E. ± 3.96).

Experiment III

There was no significant difference between serum insulin in trichinella-infected and uninfected mice over a 24-day period of study (range: 19-33 μ units/ml).

Mg% urea nitrogen in blood from infected mice was significantly less than that of uninfected from day 4 through day 24 PI (figure 4), with an average depression of 24.5% (S.E. ± 2.24).

Experiment IV

Total g body weight of trichinella-infected mice was an average of 22.5% (S.E. ± 2.14) below that of uninfected animals between days 8 and 30 PI (figure 5). Mg whole liver wet weight and dry weight per g mouse body weight in infected mice was similar to that of uninfected animals between days 4 and 30 PI (samples taken every 3rd day). Liver wet : dry weight ratio was similar in infected and uninfected mice over the same time period as above, and there was no significant difference in Mg liver protein per mg liver wet weight in infected and uninfected mice over a similar period of study (average = 208.18; S.E. ± 3.91).

In preliminary experiments it was found that although methyridine had no effect on muscle glycogen, serum insulin, or blood glucose, this drug caused a 30% increase in the liver glycogen of both infected and uninfected mice. For this reason mice in the following experiment were not given methyridine. There was no significant difference between μg liver glycogen per mg wet weight of liver tissue from infected and uninfected mice (average = 72.01; S.E. ± 1.03).

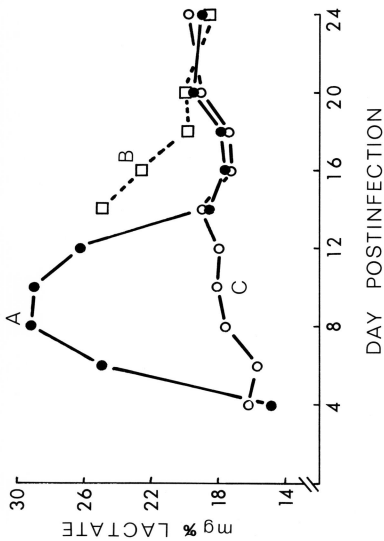

FIG. 3. Mg% BLOOD LACTATE FROM TRICHINELLA-INFECTED MICE not given methyridine (B), trichinella-infected mice given methyridine on day 11 PI (A) and uninfected animals (C). All points in line A between days 6 and 12 PI differ significantly from points on the same days in line C, and points on days 14, 16, and 18 PI in line B are significantly different from points on those days in line C. All points are the average of samples from 3 mice. p < 0.05.

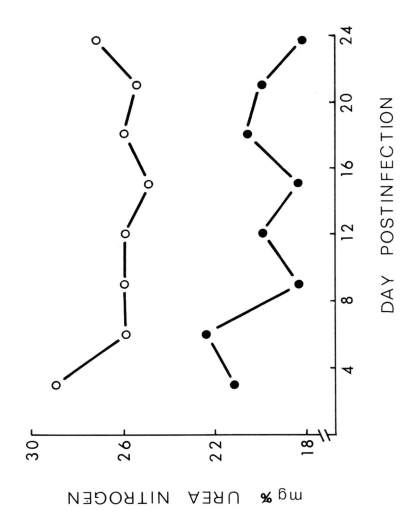

FIG. 4. Mg% UREA NITROGEN IN BLOOD from trichinella-infected (●) and uninfected mice (o). All points between days 4 and 24 PI are significantly different. All points are the average of samples from 3 mice. $p < 0.05$.

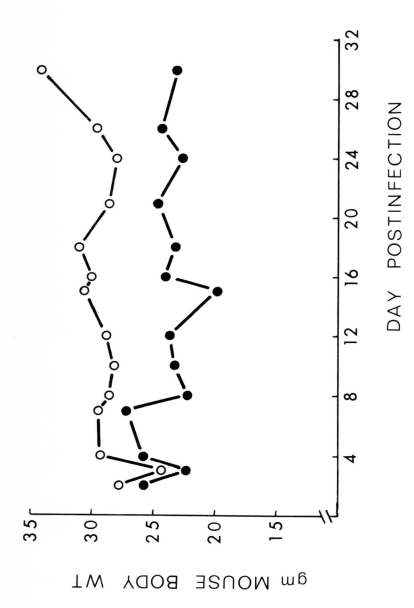

FIG. 5. TOTAL g BODY WEIGHT of trichinella-infected (●) and uninfected (o) mice. All points between days 8 and 30 PI are significantly different. All points are the average of samples from 10 mice. p < 0.05.

DISCUSSION

By the use of chemical methods, an increase in glycogen content of mouse diaphragm muscle infected with *Trichinella spiralis* was observed between days 16 and 24 PI (figure 1, line B). These findings agree with those of Zarzycki (1956) and Beckett and Boothroyd (1962), who reported an increase in histochemically demonstrable glycogen in trichinella-infected fibers between days 10 and 20 PI, followed by a decrease in infected-fiber glycogen shortly thereafter to levels similar to, or less than, uninfected fibers.

Immediately after entrance into muscle fibers, trichinella larvae undergo a three-phase growth pattern (Despommier et al., 1975). In phase III of this pattern (days 9-24 PI) muscle larvae undergo their greatest increase in size. In the present study, a dramatic rise in glycogen content of larvae occurred between days 16 and 28 PI (table 1).

Mg% blood glucose in infected mice (methyridine given on day 11 PI) was below that of uninfected animals between days 4 and 14 PI (figure 2, lines A and C), a period during which intestinal malabsorption of glucose was reported in mice infected with *T. spiralis* (days 4-16 PI; Olson and Richardson, 1968). Mg% blood glucose in infected mice not given methyridine (figure 2, line B) returned to normal levels as the number of adult trichinella present in their intestines decreased (table 2).

Mg% blood lactate in infected mice (methyridine given on day 11 PI) was above that of uninfected between days 6 and 12 PI (figure 3, lines A and C). Larsh and Race (1954) observed a mild inflammation in the intestines of mice infected with *T. spiralis*, which began about day 4 PI, entered an acute phase, which peaked around day 8 PI, and then began to subside around day 16 PI. The large numbers of inflammatory cells taking part in this cellular response may make a significant contribution to the observed increase in blood lactate from infected mice (figure 3, line A). Methyridine, given on day 11 PI, may have terminated gut inflammation (figure 3, line A) earlier than usual by removing adult worms from the intestine. On the other hand, blood lactate from infected mice not given methyridine remained elevated above that of uninfected animals until day 18 PI, when all but 17% of the initial worm dose was gone. Another significant source of lactate may be the extreme peristaltic activity of the smooth muscles of the gut. Schanbacher et al. (1976) have shown greatly enhanced gut motility associated with intestinal trichinosis. In addition, the observed decrease in mg% urea nitrogen in blood from trichinella-infected mice below that of control animals (24.5%; S.E. ± 2.24) throughout the 24-day period of study (figure 4) may imply alterations in nitrogen metabolism in the livers of trichinous mice (Hepler, 1973). Furthermore, this decrease in blood urea nitrogen may indicate a general hepatic insufficiency, including a decrease in the ability of liver tissue from trichinella-infected mice to metabolize lactate. The matter requires further investigation.

I had thought that enhanced insulin production by the pancreatic beta cells might contribute to the observed increase in trichinella-infected mouse muscle glycogen; I found no significant difference in the levels of serum insulin from infected and uninfected mice.

In agreement with the findings of Castro and Olson (1967) working with guinea pigs, I found the total mouse body weight of infected mice to be below that of control animals between days 10 and 32 PI. The ratios of liver wet or dry weight : mouse body weight, liver protein : liver wet weight, and liver glycogen : liver wet weight were similar in infected and uninfected animals.

Marked alterations occur in the biochemistry (Stewart and Read, 1972a, 1972b, 1973a, 1973b, 1974) and ultrastructure (Fasske and Themann, 1961; Ribas-Mujal and Rivera-Pomar, 1971; and Despommier, 1975) of muscle infected with *T. spiralis*. On the basis of these findings, Stewart and Read (1973b) have suggested that the damage caused by entering trichinella larvae induces muscle fibers to undergo regeneration. Shortly thereafter information from the larva redirects regeneration in host fibers to the synthesis of enzymes and structural proteins for establishing a suitable environment for the growth and development of the larva. This may include, as indicated by the present study, an enhancement of glycogenesis in infected fibers around day 12 PI. The period during which the glycogen content of trichinella larvae rises most dramatically (table 1) parallels a marked decrease in the elevated glycogen content of infected muscle between days 20 and 28 PI (figure 1, line B).

These findings lend further support to the hypothesis that trichinella-infected muscle fibers undergo redifferentiation rather than degeneration, and are metabolically active as a "nurse cell" (Purkerson and Despommier, 1974) to the trichinella larva.

ACKNOWLEDGMENT

This work was supported by grants from the NIH, U.S. Public Health Service (2 T01-AI00106 and AI-01384).

REFERENCES CITED

Barker, B. L. and W. H. Summerson
    1941 The colorimetric determination of lactic acid in biological material. Journal of Biological Chemistry 53:535-554.

Beckett, E. B. and B. Boothroyd
    1962 The histochemistry and electron microscopy of glycogen in the larva of *Trichinella spiralis* and its environment. Annals of Tropical Medicine and Parasitology 56:264-273.

Castro, G. A. and L. J. Olson
1967 Relationship between body weight and food and water intake in *Trichinella spiralis*-infected guinea pigs. Journal of Parasitology 53:589-594.

Despommier, D.
1975 Adaptive changes in muscle fibers infected with *Trichinella spiralis*. American Journal of Pathology 78:477-484.

Despommier, D., L. Aron, and L. Turgeon
1975 *Trichinella spiralis*: Growth of the intracellular (muscle) larva. Experimental Parasitology 37:108-116.

Fasske, E. and H. Themann
1961 Elektronmicroskopische Befunde an der Muskelfaser nach Trichinenbefall. Virchows Archiv: Pathologische Anatomie 334:459-474.

Gentzkow, C. J. and J. M. Masen
1942 An accurate method for the determination of blood urea nitrogen by direct Nesserlization. Journal of Biological Chemistry 143:531-537.

Hepler, O. E. (ed.)
1973 Manual of Clinical Laboratory Methods. Springfield, Ill.: Charles C. Thomas Press.

Karpiak, S. E., Z. Kozar, and M. Krzyzanowski
1963 Changes in the metabolism of skeletal muscles of guinea pigs caused by the invasion of *Trichinella spiralis*. I. Influence of the invasion on the carbohydrate metabolism of muscles. Wiadomosci Parazytologiczne 9:435-446.

Larsh, J. E., Jr., H. B. Gilchrist, and B. G. Greenberg
1952 A study of the distribution and longevity of adult *Trichinella spiralis* in immunized and non-immunized mice. J. Elisha Mitchell Scientific Society 68:1-11.

Larsh, J. E., Jr. and G. J. Race
1954 A histopathologic study of the anterior small intestine of immunized and non-immunized mice infected with *Trichinella spiralis*. Journal of Infectious Disease 94:262-272.

Lowry, O. H., N. J. Rosebrough, A. L. Farr, and R. J. Randall
1951 Protein measurement with the Folin phenol reagent. Journal of Biological Chemistry 193:265-275.

Montgomery, R.
1957 Determination of glycogen. Archives of Biochemistry and Biophysics **67**:378-386.

Olson, L. J. and J. A. Richardson
1968 Intestinal malabsorption of D-glucose in mice infected with *Trichinella spiralis*. Journal of Parasitology **54**:445-451.

Purkerson, M. and D. Despommier
1974 Fine structure of the muscle phase of *Trichinella spiralis* in the mouse. *In* Third International Conference on Trichinellosis. C. W. Kim, ed. New York: Intext.

Ribas-Mujal, D. and J. M. Rivera-Pomar
1968 Biological significance of the early structural changes in skeletal muscle fibers infected by *Trichinella spiralis*. Virchows Archiv Abteilung A: Pathologische Anatomie **345**:154-168.

Schanbacher, L. M., N. W. Weisbrodt, and G. A. Castro
1976 Parasite-induced alterations in the motility of the dog small intestine. In preparation.

Stewart, G. L. and C. P. Read
1972a Ribonucleic acid metabolism in mouse trichinosis. Journal of Parasitology **58**:252-256.

1972b Some aspects of cyst synthesis in mouse trichinosis. Journal of Parasitology **58**:1061-1064.

1973a Deoxyribonucleic acid metabolism in mouse trichinosis. Journal of Parasitology **59**:264-267.

1973b Changes in RNA in mouse trichinosis. Journal of Parasitology **59**:997-1005.

1974 Studies on biochemical pathology in trichinosis. I. Changes in myoglobin, free creatine, phosphocreatine; and two protein fractions of mouse diaphragm muscle. Journal of Parasitology **60**:996-1000.

Zarzycki, J.
1956 Histologic investigation on the glycogen content in striated muscle infested with Trichinellae. Med. Weteryn. **12**:328-332.

ABSORPTION KINETICS OF SOME PURINES, PYRIMIDINES, AND NUCLEOSIDES IN *TAENIA CRASSICEPS* LARVAE

by Gary L. Uglem and Michael G. Levy

ABSTRACT

In vitro kinetic studies revealed that *Taenia crassiceps* larvae absorb purine bases (adenine and hypoxanthine) and nucleosides (adenosine and uridine) by mediated transport. Pyrimidines (thymine, uracil, and cytosine) are absorbed primarily by diffusion. The presence of hypoxanthine stimulated adenosine uptake, an effect which was enhanced by preloading the worms with hypoxanthine. Furthermore, of the compounds tested, hypoxanthine alone displayed sigmoid absorption kinetics, suggesting an allosteric mechanism. This mechanism was repressed by preloading the worms with hypoxanthine, however, as well as by decreasing the incubation time from 2 minutes to 30 seconds. Under the latter experimental conditions hypoxanthine uptake appeared to occur by diffusion alone. The results suggest that the allosteric effects of hypoxanthine in this organism may involve an effect of hypoxanthine on metabolism and/or the adsorption of hypoxanthine and adenosine on cytoplasmic binding sites.

INTRODUCTION

Radiographic experiments have shown that certain parasitic helminths rapidly incorporate absorbed purines and pyrimidines into nucleic acids (Prescott and Voge, 1959; Dvorak and Jones, 1963). Kinetic studies of the membrane transport of these bases are scarce, however, and are limited to studies of adult worms.

In this paper we describe two systems in *Taenia crassiceps* larvae for transporting nitrogenous bases and nucleosides. The allosteric nature of hypoxanthine absorption and the stimulatory effect of hypoxanthine on nucleoside uptake are also reported.

Gary Uglem is Assistant Professor of Biology at the University of Kentucky. Michael Levy is Postdoctoral Fellow at Harvard University School of Public Health.

MATERIALS AND METHODS

Krebs-Ringer saline solution containing 25 mM tris(hydroxymethyl)-aminomethane-maleate buffer at pH 7.4 (KRT of Read et al., 1963) was the solution used in all washings, preincubations, and incubations. The labeled compounds used were adenine-8-^{14}C, thymine-2-^{14}C, adenosine-^{3}H(G), uridine-^{3}H(G) (Amersham Searle), uracil-2-^{14}C, hypoxanthine-8-^{14}C, and cytosine-2-^{14}C (New England Nuclear). Unlabeled compounds were obtained from Sigma and Calbiochem. Incubation media consisted of 5 ml of KRT containing a single labeled substrate with or without the appropriate un-labeled inhibitor.

Methods for maintaining laboratory infections of *Taenia crassiceps* larvae have been described previously (Pappas et al., 1973). Larvae were flushed from the body cavity of an infected mouse and washed several times. The small solid larvae were collected and distributed into random groups (40 to 80 mg wet wt/group). Each group of larvae was preincubated for 15 minutes (37° C) before incubation with the labeled substrate. Unless indicated other-wise, all incubations were for 2 minutes. To stop the incubations, larvae were washed briefly three times, blotted dry, and placed in 3 ml of 70% ethanol. After 24 hours the larvae were removed, dried for 24 hours at 95° C, and weighed. Radioactivity in aliquots of the ethanol extracts was determined using a scintillation spectrometer (Beckman). Data were compared using the Student's "t" test. The kinetic parameters (V_{max} and K_t) were calculated using the Lineweaver-Burk analysis.

RESULTS

Taenia crassiceps larvae absorbed ^{14}C-thymine and -uracil in direct rela-tion to substrate concentration; the worms were somewhat more permeable to thymine than to uracil as indicated by the diffusion rates (figure 1). Furthermore, at a low fixed substrate concentration (0.05 mM) the influx rates of these pyrimidines were not significantly inhibited by the addition of unlabeled molecules of the same species (figure 1, inset). Unlabeled cytosine (10 mM) inhibited the influx of 0.05 mM ^{14}C-cytosine, indicating the presence of a small mediated component (not shown). A kinetic analysis of cytosine transport would be difficult, however, because of the large rate of entry by diffusion. Thus, the main route of entry for pyrimidines is by diffusion.

FIG. 1 (OPPOSITE). UPTAKE OF ^{14}C-THYMINE (●) AND -URACIL (o) by *Taenia crassiceps* larvae. V = μmoles absorbed/g ethanol extracted dry wt/2 min. [S] = substrate concentration in mM. INSET: Uptake of 0.05 mM ^{14}C-thymine (●) and -uracil (o) in the presence of unlabeled inhibitor of the same molecular species. [I]=inhibitor concentration in mM. Each point is the average of three samples.

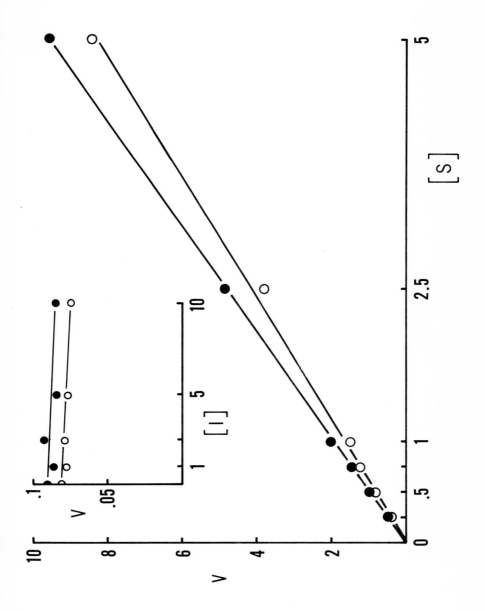

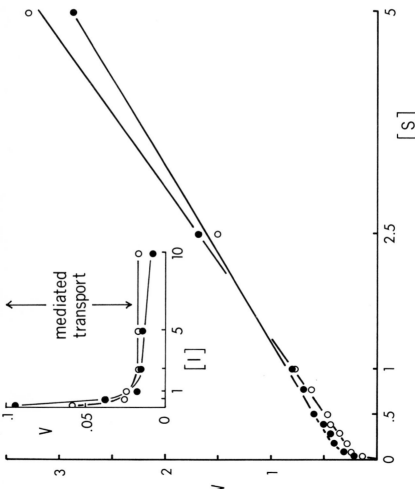

FIG. 2. UPTAKE OF ^{3}H-URIDINE (●) AND -ADENOSINE (o) by *Taenia crassiceps* larvae. V = μmoles absorbed/g ethanol extracted dry wt/2 min. [S] = substrate concentration in mM. INSET: Uptake of 0.05 mM ^{3}H-uridine (●) and -adenosine (o) in the presence of unlabeled inhibitor of the same molecular species. [I] = inhibitor concentration in mM. Each point is the average of three samples.

[3]H-adenosine and -uridine entered larvae by a combination of mediated transport and diffusion (figure 2). Although the larvae appeared to be freely permeable to both compounds, mediated transport accounts for most of the uptake at 0.05 mM substrate concentration (figure 2, inset). Because of the small number of substrate concentrations tested below 0.2 mM, an accurate kinetic analysis could not be made. Nevertheless, the kinetic parameters for these nucleosides can be estimated by looking at the data in figure 2. The V_{max} is between 0.17 and 0.3 μmoles/g dry wt/2 min, while the K_t is near 0.05 mM. Uridine transport appears to have a higher V_{max}, but a somewhat lower rate of entry by diffusion.

The effects of some purines and nucleotides on adenosine and uridine transport were observed (table 1). Hypoxanthine (6 mM) and adenine (10 mM) stimulated adenosine transport, while uridine transport was little affected by the compounds. If the worms were preincubated in 6 mM hypoxanthine for 15 minutes before incubation, hypoxanthine further stimulated adenosine transport. A maximum rate of stimulation was observed even when hypoxanthine was absent during incubation with labeled adenosine. AMP and ribose inhibited transport of these nucleosides.

TABLE 1

EFFECTS OF VARIOUS COMPOUNDS ON THE UPTAKE OF 0.05 mM
[3]H-ADENOSINE AND -URIDINE BY *TAENIA CRASSICEPS*
LARVAE IN 2 MINUTE INCUBATIONS

The numbers in parentheses represent the percentage of inhibition (I) or stimulation (S). Uptake rates are expressed as μmoles/g ethanol extracted dry wt/hr. Each value is the mean $\pm$ SE of three replicates. Values are uncorrected for diffusion.

INHIBITOR	SUBSTRATE			
(10 mM)	Adenosine		Uridine	
None	1.62 ± 0.16	—	5.70 ± 0.01	—
Adenosine	0.35 ± 0.02	(78% I)	0.44 ± 0.05	(92% I)
Uridine	0.39 ± 0.06	(76% I)	0.35 ± 0.05	(94% I)
Thymine	1.11 ± 0.49	(0)	5.38 ± 0.63	(0)
Uracil	1.58 ± 0.10	(0)	4.88 ± 0.38	(0)
Hypoxanthine (6 mM)	3.14 ± 0.25	(94% S)	5.89 ± 0.11	(0)
Adenine	3.80 ± 0.72	(134% S)	6.39 ± 0.13	(12% S)
AMP	0.60 ± 0.32	(63% I)	1.97 ± 0.24	(46% I)
Ribose	0.97 ± 0.33	(40% I)	4.24 ± 0.26	(26% I)

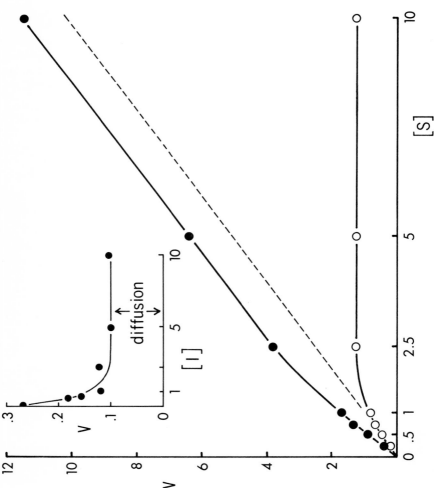

FIG. 3. UPTAKE OF ^{14}C-ADENINE by *Taenia crassiceps* larvae. $V = \mu$moles absorbed/g ethanol extracted dry wt/2 min. [S] = substrate concentration in mM. Dashed line represents the diffusion rate. Mediated transport (o) was calculated by subtracting diffusion from the observed rates (●). INSET: Uptake of 0.05 mM ^{14}C-adenine in the presence of unlabeled adenine as inhibitor ([I] = mM). Each point is the average of three samples.

Two mediated systems appear to operate in the absorption of purines. The graph of ^{14}C-adenine transport exhibited typical saturation kinetics (figure 3), with a rate of diffusion similar to that of the pyrimidines. Both the calculated V_{max} (1.51 μmoles/g dry wt/2 min) and K_t (1.47 mM) were higher than those estimated for adenosine uptake. The rates of diffusion as determined from the V *vs.* [S] plot (figure 3) and the V *vs.* [I] plot (figure 3, inset) are in agreement. The fact that 10 mM unlabeled adenosine, uridine, or AMP did not affect the influx rate of 0.1 mM adenine indicates that adenine is transported at a separate site.

In contrast to adenine uptake, the rate of hypoxanthine uptake showed a sigmoid absorption pattern with respect to substrate concentration (figure 4), indicating the presence of an allosteric mechanism. Uptake appeared to be linear between 0.1 and 0.5 mM, was non-linear from 0.5 to 2.5 mM, and was again linear at concentrations greater than 2.5 mM. Unlabeled hypoxanthine (6 mM) inhibited the uptake of 0.05 mM ^{14}C-hypoxanthine by 21% (figure 5, solid circles).

The above absorption pattern for hypoxanthine was changed by decreasing the incubation time. In 30 second incubations, the uptake rate was more linear with respect to increasing substrate concentrations (figure 6). Furthermore, the uptake of 0.05 mM ^{14}C-hypoxanthine in 30 second incubations was not inhibited by unlabeled hypoxanthine, indicating that uptake is solely by diffusion (figure 6, inset). These patterns of absorption (i.e., linear V *vs.* [S] curve, and lack of inhibition in a V *vs.* [I] curve) could also be produced in 2 minute incubations by first preincubating the worms in 6 mM unlabeled hypoxanthine for 15 minutes before incubation with the substrate (figure 4, open circles; figure 5, open circles).

DISCUSSION

Previous studies have suggested that purines and pyrimidines (MacInnis et al., 1965; MacInnis and Ridley, 1969; Pappas et al., 1973) and nucleosides (Page and MacInnis, 1975) move across the plasma membrane of *H. diminuta* by mediated transport. While several specific transport systems or loci have been identified for these compounds, the data are often difficult to interpret when the kinetic plots result in sigmoid-shaped curves. Furthermore, the presence of one compound might increase the rate of transport of another. To explain these apparent allosteric effects, the presence of two sites per locus has been postulated such that activation of the allosteric site increases the rate of translocation via the transport site. Although the molecular specificities of the allosteric systems in *H. diminuta* have been well defined, the significance of such systems is not clear.

We had hoped to determine the presence of mediated transport systems for nitrogenous bases in *T. crassiceps* larvae. The experiments demonstrated the

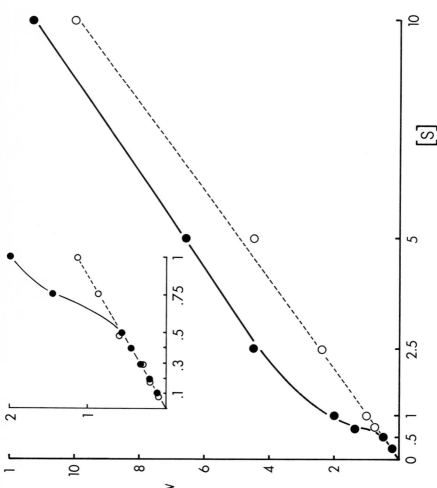

FIG. 4. UPTAKE OF ^{14}C-HYPOXAN-THINE (●) by *Taenia crassiceps* larvae. V = μmoles absorbed/g ethanol extracted dry wt/2 min. [S] = substrate concentration in mM. Some groups of worms (o) were preincubated with 6 mM unlabeled hypoxanthine for 15 minutes prior to incubation with the labeled substrate. Some of the points between 0.1 and 1.0 mM that are omitted for clarity are given in the inset. Each point is the average of three samples.

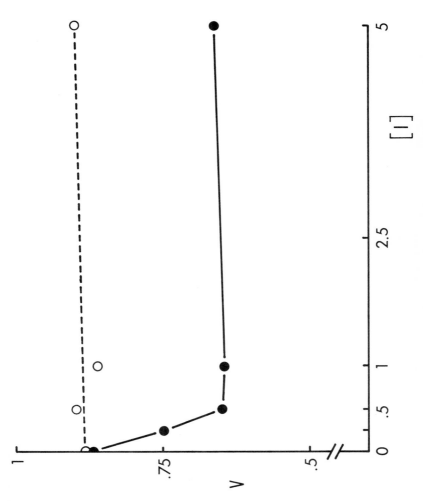

FIG. 5. UPTAKE OF 0.05 mM ^{14}C-HYPOXANTHINE (●) by *Taenia crassiceps* larvae in the presence of unlabeled hypoxanthine as inhibitor ([I] = mM). V = μmoles absorbed/g ethanol extracted dry wt/2 min. Some worms (o) were preincubated with 6 mM unlabeled hypoxanthine for 15 minutes before incubation with the substrate. Each point is the average of three samples.

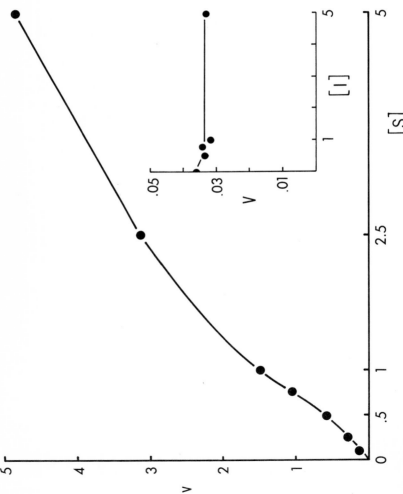

FIG. 6. UPTAKE OF ^{14}C-HYPOXAN-THINE by *Taenia crassiceps* larvae in 30 second incubations. $V = \mu$moles absorbed/g ethanol extracted dry wt/30 sec. [S] = substrate concentration in mM. INSET: Uptake of 0.05 mM ^{14}C-hypoxanthine in the presence of unlabeled hypoxanthine as inhibitor ([I] = mM). Each point is the average of three samples.

presence of such systems for purines and nucleosides. No allosteric system, however, such as exists in *H. diminuta*, was detected for pyrimidines in *T. crassiceps*. In fact, thymine and uracil appeared to be absorbed by diffusion alone. Until larvae of *H. diminuta* (or adults of *T. crassiceps*) are examined, we can only speculate whether the capacity for pyrimidine transport is characteristic of certain species or is a biochemical phenomenon found in larval but not adult cestodes.

While hypoxanthine stimulates adenosine uptake in both species of tapeworms, the mechanisms for absorbing hypoxanthine are different. In *H. diminuta* a plot of the velocity of uptake of hypoxanthine versus hypoxanthine concentration yields a typical saturation curve (MacInnis et al., 1965). Such a plot in *T. crassiceps* is distinctly sigmoidal. Furthermore, this effect in *T. crassiceps* is repressed by decreasing the incubation time to 30 seconds. Since binding of hypoxanthine to an allosteric site may not be significant in 30 seconds, some worms were preloaded first with hypoxanthine in an attempt to "prime" the system. Surprisingly, not only did this treatment fail to restore transport activity in 30 second incubations, but the allosteric pattern of uptake in 2 minutes was also repressed; the uptake of 0.05 mM labeled hypoxanthine was not inhibited by unlabeled hypoxanthine, and the V *versus* [S] plot was linear. On the other hand, hypoxanthine-loaded worms transported adenosine at higher rates even when hypoxanthine was not present in the incubation medium.

The interactions of purines and nucleosides in *T. crassiceps* are complex and difficult to explain by reference to membrane phenomena alone. If the allosteric effects of hypoxanthine prove to be due to the adsorption of hypoxanthine on sites within the cell (and if they in this way stimulate co-adsorption and/or metabolism of adenosine), then perhaps a more inclusive concept of transport and accumulation of nutrients in parasitic helminths is needed.

ACKNOWLEDGMENT

This study was supported by Biomedical Sciences Support Grant No. 5 S05 R R07115-07 from the General Research Support Branch, Division of Research Resources, Bureau of Health Professions Education and Manpower Training, National Institutes of Health.

REFERENCES CITED

Dvorak, J. A. and A. W. Jones
   1963 *In vivo* incorporation of tritiated cytidine and tritiated thymidine by the cestode, *Hymenolepis microstoma*. Experimental Parasitology **14**:316-322.

MacInnis, A. J., F. M. Fisher, and C. P. Read
1965 Membrane transport of purines and pyrimidines in a cestode. Journal of Parasitology 51:260-267.

MacInnis, A. J. and R. K. Ridley
1969 The molecular configuration of pyrimidines that causes allosteric activation of uracil transport in *Hymenolepis diminuta*. Journal of Parasitology 55:1134-1140.

Page, C. R., III and A. J. MacInnis
1975 Characterization of nucleoside transport in hymenolepidid cestodes. Journal of Parasitology 61:281-290.

Pappas, P. W., G. L. Uglem, and C. P. Read
1973 The influx of purines and pyrimidines across the brush border of *Hymenolepis diminuta*. Parasitology 66:525-538.

Prescott, D. M. and M. Voge
1959 Autoradiographic study of the synthesis of ribonucleic acid in cysticercoids of *Hymenolepis diminuta*. Journal of Parasitology 45:587-590.

Read, C. P., A. H. Rothman, and J. E. Simmons
1963 Studies on membrane transport, with special reference to parasite-host integration. Annals of the New York Academy of Sciences 113:154-205.